T0395443

VOLUME TWO HUNDRED AND THIRTY FIVE

ADVANCES IN IMAGING AND ELECTRON PHYSICS

VOLUME TWO HUNDRED AND THIRTY FIVE

ADVANCES IN IMAGING AND ELECTRON PHYSICS

Edited by

MARTIN HŸTCH
CEMES-CNRS
Toulouse, France

Cover photo credit: Maxwell's equations for an inhomogeneous medium with sources, from Chapter 1, Equation 5.

Academic Press is an imprint of Elsevier
125 London Wall, London, EC2Y 5AS, United Kingdom
50 Hampshire Street, 5th Floor, Cambridge, MA 02139, United States

ISBN: 978-0-443-42831-9
ISSN: 1076-5670

For information on all Academic Press publications
visit our website at https://www.elsevier.com/books-and-journals

Publisher: Zoe Kruze
Acquisition Editor: Jason Mitchell
Editorial Project Manager: Palash Sharma
Production Project Manager: James Selvam
Cover Designer: Arumugam Kothandan

Typeset by MPS Limited, India

Contents

Contributors

Román Castañeda
Physics Department, Universidad Nacional de Colombia Sede Medellín, Medellín, Colombia

Ramaswamy Jagannathan
The Institute of Mathematical Sciences, Central Institutes of Technology (CIT) Campus, Tharamani, Chennai, India

Sameen Ahmed Khan
Department of Mathematics and Sciences, College of Arts and Applied Sciences (CAAS), Dhofar University, Salalah, Sultanate of Oman

Preface

Peter Hawkes was a renowned theoretician in the field of electron optics. As editor-in-chief of AIEP, he encouraged the publication of new theoretical developments and I intend to continue this endeavor. It is fitting then that both chapters in this volume 235 of AIEP are from authors whose work he knew well and who have regularly contributed to AIEP in the past. The first lengthy contribution is from Sameen Ahmed Khan and Ramaswamy Jagannathan who develop a matrix formulation of wave optics that explicitly includes polarization. In a similar spirit to their description that fully accounts for the spin of the electron in electron optical design (AIEP volume 229), they continue to reexamine the mathematical basis of optics, charged particle or light optics, to place the theory on a rigorous basis with practical applications always in mind. The second chapter is from Román Castañeda on quantum theory for interference for single matter particles or photons. This is an extension and generalization of his previous contribution in AIEP volume 229. Particle interference remains a subject where everyday intuition collides with conventional quantum mechanical interpretation. It is welcome that physicists continue to explore the reconciliation of the two.

As always, we thank our contributors for their efforts to make very specialized material accessible to readers from other fields.

Martin Hÿtch
Toulouse

CHAPTER ONE

A matrix formalism of the Maxwell vector wave optics including polarization☆

Sameen Ahmed Khan[a,*] and Ramaswamy Jagannathan[b,1]
[a]Department of Mathematics and Sciences, College of Arts and Applied Sciences (CAAS), Dhofar University, Salalah, Sultanate of Oman
[b]The Institute of Mathematical Sciences, Central Institutes of Technology (CIT) Campus, Tharamani, Chennai, India
*Corresponding author. e-mail address: rohelakhan@yahoo.com

Contents

☆ *Dedicated to the memory of Dr. Peter William Hawkes whose constant support was part of our scientific lives since 1990s*
[1] Retired

Advances in Imaging and Electron Physics, Volume 235
ISSN 1076-5670, https://doi.org/10.1016/bs.aiep.2025.06.002

Abstract

A matrix formalism of the Maxwell vector wave optics including polarization is presented based on a new matrix representation of the Maxwell equations for a linear inhomogeneous medium. This formalism results in an exact Maxwell vector wave optical Hamiltonian without any assumptions on the form of the varying refractive index. The scalar wave optics subduced by the Maxwell vector wave optics leads to small wavelength-dependent corrections to the Helmholtz scalar wave optics obtained by the traditional and nontraditional methods. As an example, we consider an axially symmetric graded-index medium and derive the scalar wave optical aberration coefficients which include small wavelength-dependent corrections to the traditional expressions. As for polarization, besides deriving the Mukunda-Simon-Sudarshan matrix substitution rule (MSS rule) for transition from the Helmholtz scalar wave optics to the Maxwell vector wave optics, we also express it in terms of a second order matrix differential operator. The power of the new formalism is demonstrated by applying it to study cross polarization in seven specific types of beams like the Gaussian, Bessel, and Airy beams. This matrix formalism of the Maxwell vector wave optics extends Hamilton's optical-mechanical analogy beyond the ray optics and the scalar wave optics.

1. Introduction

The geometrical, or ray, optics can be derived fully from Fermat's principle of least time (see, e.g., Buchdahl, 1993; Lakshminarayanan et al., 2002). This geometric approach leads to a Hamiltonian description of an optical system, with the optical Hamiltonian prescribed to the desired degree of accuracy, explaining the paraxial behaviour and aberrations of the sytem to any order. The Lie algebraic formulation of ray optics deserves a special mention and is just a few decades old (Dragt, 1982, 1988; Dragt & Forest, 1986; Dragt et al., 1986, 1988; Lakshminarayanan et al., 1998; Rangarajan & Sachidanand, 1997;

Rangarajan & Sridharan, 2010; Rangarajan et al., 1990). Lie algebraic approach has clarified several issues and enables a precise way of enumerating and classifying aberrations in terms of polynomials derived from the optical Hamiltonian. A complete description of all aspects of optics is possible only on the basis of the Maxwell equations of electromagnetism (see, e.g., Born & Wolf, 1999). The Maxwell equations, though linear, are coupled and constrained. So, mathematically it is not easy to construct optical formalisms directly from the Maxwell equations. A widely used formalism is the scalar wave optics based on the Helmholtz equation derived from the Maxwell equations. However, there are compromises: the Maxwell equations are of first order whereas the Helmholtz equation is of second order and approximate. The Helmholtz equation provides a Hamiltonian approach to scalar wave optics which goes over to the ray optics in the short wavelength limit. On a closer examination, the procedure of deducing the scalar wave optical Hamiltonian from the Helmholtz equation is not found to be mathematically rigorous enough. This deduction can be done in an alternate way adopting the techniques of quantum mechanics used for studying the Klein-Gordon equation. This is possible because the Klein-Gordon equation and the Helmholtz equation have a strikingly similar mathematical structure though they correspond to different physical systems. This results in the non-traditional formalism of the Helmholtz scalar wave optics (Khan, 2005b, 2016b, 2018a, 2018c; Khan et al., 2002). In the non-traditional formalism of the Helmholtz optics the Hamiltonian has wavelength-dependent correction terms. Consequently, the paraxial behaviour and aberrations to all orders get modified by wavelength-dependent corrections. But the fact remains that the basic equation of this theory, the Helmholtz equation, is an approximation to the Maxwell equations (see Appendix A).

Importantly, the study of polarization has to be necessarily based on the Maxwell equations (see, e.g., Born & Wolf, 1999). So, it is desirable to have a formalism of the Maxwell vector wave optics which can explain all aspects of optics including polarization. In this direction, the pioneering work of Sudarshan et al. has established a formalism of Fourier optics for the Maxwell field, based on group theoretical considerations of the Maxwell equations, enabling an unambiguous transition from the scalar paraxial wave optics to the Maxwell vector Fourier optics (Mukunda et al., 1983, 1985a, 1985b; Simon et al., 1986, 1987; Sudarshan et al., 1983). As the Maxwell equations are coupled and constrained, it is not easy to deal with them directly. Seeking an alternative approach to the problem, a matrix representation of the Maxwell equations for a linear inhomogeneous medium was derived by one of us (Khan, 2005a), heuristically, starting with the equations obeyed by the Riemann-Silberstein-Weber (RSW) vector. Trying to build a matrix formalism of the Maxwell vector wave optics, using this representation (Khan, 2005a), has led us recently to a new matrix representation of the Maxwell equations for a linear inhomogeneous medium (Khan & Jagannathan, 2024c). This new matrix representation has been derived starting *ab initio* with the Maxwell equations, following the work of Bocker and Frieden (1993, 2018) on the matrix representtion of the Maxwell equations for the vacuum. Though our new representation is equivalent to the earlier one (Khan, 2005a), it has certain advantages over the earlier one and is more suitable for building the desired matrix formalism of the Maxwell vector wave optics.

The present Chapter elaborates on the matrix formalism of the Maxwell vector wave optics including polarization built on our recent work (Khan & Jagannathan, 2024c). We present our matrix formalism primarily in terms of the refractive index and the impedance of the medium. As has been shown by us (Khan & Jagannathan, 2024c), in this formalism the RSW vector emerges as a natural basis rather than being used as an *ad hoc* building block to start with. The new matrix formalism presented here readily leads to an exact Maxwell vector wave optical Hamiltonian. This Maxwell vector wave optical Hamiltonian has exactly the same mathematical structure as Dirac's relativistic quantum Hamiltonian for electron. This ensures that the standard techniques of relativistic quantum mechanics of electron can be adopted to treat the Maxwell vector wave optics. The chief technique in this case is the Foldy-Wouthuysen (FW) expansion of the Hamiltonian. From the onset, let us bear in mind that the six-dimensional electromagnetic field and the four-dimensional Dirac spinor are two entirely different physical entities.

The Foldy-Wouthuysen transformation (FWT) technique was originally developed for the Dirac equation for electron (Acharya & Sudarshan, 1960; Bjorken & Drell, 1994; Case, 1954; Costella & McKellar, 1995; Foldy, 1952; Foldy & Wouthuysen, 1950; Greiner, 2000; Khan, 2006, 2008; Orris & Wurmser, 1995; Osche, 1977; Silenko, 2016). But it turns out that it can be applied to a certain wider class of matrix evolution equations. Our new matrix representation of the Maxwell equations falls within this special class of matrix evolution equations (Khan, 2006, 2008, 2017a). The exact similarities in the underlying mathematical structures of the two systems ensure that we can adapt the powerful machinery of the FW theory. Using the mathematical techniques of quantum theory in the study of classical optical systems has been called quantum methodology. The use of such quantum methodology (particularly, the Foldy-Wouthuysen-like transformation technique) has served as a powerful analytical tool leading to well established results, and even in predicting new effects in the Maxwell vector wave optics

(Khan, 2010, 2014a, 2016c, 2016d, 2017a, 2017b, 2017c, 2017e, 2018b, 2018c, 2020, 2021, 2022, 2023a, 2023b, 2023c, 2023d, 2024). The optical Hamiltonian of any system in the matrix formalism of the Maxwell vector wave optics leads to wavelength-dependent corrections to the results on behaviour of the system obtained on the basis of traditional and non-traditional Helmholtz scalar wave optical treatments. This is exactly analogous to the situation in charged particle beam optics. Quantum theory of charged spinor particle beam optics leads to modifications in the quantum theory of charged scalar particle beam optics depending on the de Broglie wavelength of the particle (Jagannathan & Khan, 1996, 2019). Thus, the matrix formalism of the Maxwell vector wave optics extends Hamilton's optical-mechanical analogy beyond the ray optics and the scalar wave optics.

This Chapter includes the following topics with examples.

1. *Medium with constant refractive index and constant impedance*:
   This constitutes an ideal system and can be analyzed exactly. The exact result is used to illustrate the machinery of the Foldy-Wouthuysen-like (FW-like) series expansion and compare it with the exact result.
2. *Axially symmetric graded-index medium*:
   Graded-index optical devices have wide applications (see, e.g., Gomez-Reino et al., 2002). We work out the theory of an axially symmetric graded-index medium and consider the example of a graded-index lens (GRIN lens). Two points are worth mentioning about this system. In the traditional scalar wave theory, one obtains six third-order aberrations. In a non-traditional treatment of the Helmholtz scalar wave optics these six aberration coefficients are modified by wavelength-dependent corrections (Khan, 2005b, 2016b, 2018a, 2018c; Khan et al., 2002). Scalar wave optics subduced by the matrix formalism of the Maxwell vector wave optics (see Section 5) leads to a tiny wavelength-dependent image rotation in a GRIN lens. We have derived its magnitude (see (181) in Section 5.2.2). Here, the term 'subduced' is borrowed from group theory and its definition is presented in Section 5.2.1. Axial symmetry permits exactly nine third-order aberrations. Six of these are obtained in the Helmholtz scalar wave theory and get modified by wavelength-dependent corrections in a non-traditional treatment of the Helmholtz scalar wave optics. The scalar wave optics subduced by the matrix formalism of the Maxwell vector wave optics includes the remaining three third order aberrations as pure wavelength-dependent effects. We have obtained the expressions for all the nine aberration coefficients which explicitly have the wavelength-dependent

corrections to the standard expressions (see (201) in Section 5.2.3). This is analogous to the existence of nine aberrations and image rotation in the case of an axially symmetric magnetic lens in classical charged particle beam optics. The quantum treatment of this system modifies the classical aberration coefficients by the de Broglie wavelength-dependent correction terms (Jagannathan & Khan, 1996, 2019; Khan, 1997).

3. *Mukunda-Simon-Sudarshan matrix substitution rule for transition from the Helmholtz scalar wave optics to the Maxwell vector wave optics*:

   Understanding and manipulating light beams and polarization are the foundations of optical technologies (Otte, 2020). In the 1980s, a systematic procedure for transition from the Helmholtz scalar wave optics to the Maxwell vector wave optics was developed by analyzing the relativistic symmetry of the Maxwell equations. This procedure derived using group theoretical techniques is now known as the Mukunda-Simon-Sudarshan matrix substitution rule, or simply the MSS substitution rule, or just the MSS rule for transition from the Helmholtz scalar wave optics to the Maxwell vector wave optics (Mukunda et al., 1983, 1985a, 1985b; Simon et al., 1986, 1987; Sudarshan et al., 1983). Besides many other applications, the MSS rule is useful for obtaining the Maxwell beams which are solutions of the Maxwell equations. The remarkable feature of this application of the MSS rule is that, it makes use of the widely available and easy to derive solutions of the Helmholtz scalar equation, rather than solving the more difficult system of Maxwell's equations. In fact, even the Helmholtz scalar wave equation is solved using the parabolic approximation with appropriate physical reasoning (see, e.g., Lakshminarayanan & Varadharajan, 2015). The elegant MSS-rule emerges from the paraxial approximation of the matrix formalism presented here.

4. *A matrix differential operator form of the MSS rule*:

   It is further shown that the MSS rule is equivalent to the action of a matrix differential operator. It is seen that this matrix differential operator form of the MSS rule is much easier to use than the original matrix substitution rule (Khan, 2023b). We demonstrate the action of the MSS rule in its both forms, namely, the matrix substitution form and the equivalent differential operator form. We use the MSS rule to study the cross polarization in the following seven light beams: Gaussian beams, anisotropic Gaussian beams, Hermite-Gaussian beams, Bessel beams, Bessel-Gaussian beams, Airy beams and the anisotropic Airy beams.

The layout of this Chapter is as follows. In Section 2, we develop the matrix formalism of the Maxwell vector wave optics. In Section 3 we develop the FW-like successive approximation technique for beam optical systems. In Section 4, we describe briefly the transition to the Helmholtz scalar wave optics and ray optics. In Section 5 we apply the matrix formalism of the Maxwell vector wave optics to two specific systems: (a) medium with a constant refractive index and constant impedance; (b) an axially symmetric graded-index medium. In Section 6 we derive the Mukunda-Simon-Sudarshan matrix substitution rule for transition from the Helmholtz scalar wave optics to the Maxwell vector wave optics. We also show that the MSS rule can be implemented in a much easier way by the action of a matrix differential operator. In Section 7 we demonstrate the application of the MSS-rule to study the cross polarization in seven specific light beams. In Section 8 we conclude with some remarks. There are four Appendices. Appendix A has the derivation of the generalized wave equation for the electric field $\boldsymbol{E}$ in an inhomogeneous medium mentioned in Section 2. This derivation is done using the present matrix formalism. Appendix B spells out the main properties of the RSW vector. Appendix C has a brief outline of different formulations of the ray optics and the Helmholtz scalar wave optics. Our approach to Maxwell beams requires the solutions of the Helmholtz scalar wave equation. The required solutions for a diverse set of seven light beams are presented in Appendix D. Lastly, we have an extensive bibliography.

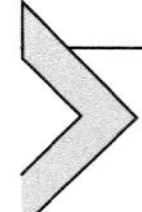

2. A matrix formalism of the Maxwell vector wave optics

Matrix, or Dirac-like spinor, representation of Maxwell's equations for the electromagnetic field, mostly in vacuum, has a long history starting with the works of Laporte and Uhlenbeck (1931), Majorana (1931; Unpublished Notes, see Mignani et al., 1974), Oppenheimer (1931), and others (see Barnett, 2014; Belkovich. & Kogan, 2016; Bialynicki-Birula, 1994, 1996a, 1996b; Bialynicki-Birula & Bialynicka-Birula, 2013; Bogush et al., 2009; Dvoeglazov, 1993; Edmonds, 1975; Esposito, 1998; Giannetto, 1985; Good, 1957; Inskeep, 1988; Ivezć, 2006; Jestädt et al., 2019; Kisel et al., 2011; Kulyabov, 2016; Kulyabov et al., 2017; Livadiotis, 2018; Mingjie et al., 2020; Mohr, 2010; Moses, 1959; Ohmura, 1956; Ovsiyuk et al., 2013; Red'kov et al., 2012; Sachs & Schwebel, 1962;

Wang, 2015, and references therein). Many of these matrix representations are six dimensional and often given in terms of a pair of matrix equations involving the Riemann-Silberstein (Silberstein, 1907a, 1907b), or the Riemann-Silberstein-Weber (RSW) vector (Kiesslinga & Tahvildar-Zadehb, 2018; Sebens, 2019). An eight dimensional matrix representation of the Maxwell equations, not based on the RSW vector, was also known (Bocker & Frieden, 1993). A new eight dimensional matrix representation of the Maxwell equations for a linear inhomogeneous medium based on the RSW vector was derived by one of us (Khan, 2005a) and it has been found to be useful in studying several problems of electromagnetic theory (Mehrafarin & Balajany, 2010; Ram et al., 2021; Vahala et al., 2020a, 2020b, 2020c, 2021a, 2021b, 2022a, Vahala et al., 2022), and to understand several aspects of the optics of the Maxwell vector wave beams including polarization (Khan, 2006, 2008, 2010, 2014a, 2016c, 2016d, 2017a, 2017b, 2017c, 2017e, 2020, 2021, 2022, 2023a, 2023b, 2023c, 2023d, 2024). Following (Bocker & Frieden, 1993, 2018), this representation (Khan, 2005a) has been developed further by us (Khan & Jagannathan, 2024c). The RSW vector turns out to be the natural basis for this new representation.

Maxwell equations for a linear inhomogeneous medium contain two physical parameters, namely, the permittivity $\epsilon(\boldsymbol{r}, t)$ and the permeability $\mu(\boldsymbol{r}, t)$ of the medium. Speed of light in a medium is given by

$$v(\boldsymbol{r}, t) = \frac{1}{\sqrt{\epsilon(\boldsymbol{r}, t)\mu(\boldsymbol{r}, t)}}. \tag{1}$$

With ϵ_0 and μ_0 denoting the constant permittivity and permeability of vacuum, respectively, the constant speed of light in vacuum is

$$c = \frac{1}{\sqrt{\epsilon_0 \mu_0}}. \tag{2}$$

In optics, the chief quantity is the refractive index of the medium defined by

$$n(\boldsymbol{r}, t) = \frac{c}{v(\boldsymbol{r}, t)} = c\sqrt{\epsilon(\boldsymbol{r}, t)\mu(\boldsymbol{r}, t)}. \tag{3}$$

The refractive index of vacuum is 1 by definition. The spatial and temporal variations of permittivity and permeability of a medium are responsible for the variation of the refractive index of the medium. Another quantity of interest is the impedance of the medium defined by

$$\eta(\boldsymbol{r},\, t) = \sqrt{\frac{\mu(\boldsymbol{r},\, t)}{\epsilon(\boldsymbol{r},\, t)}}\,. \tag{4}$$

Maxwell's equations for an inhomogeneous medium with sources are given by (see, e.g., Born & Wolf, 1999; Jackson, 1998; Panofsky & Phillips, 1962):

$$\begin{aligned} \frac{\partial \boldsymbol{D}(\boldsymbol{r},t)}{\partial t} &= \nabla \times \boldsymbol{H}(\boldsymbol{r},\, t) - \boldsymbol{J}, \\ \nabla \cdot \boldsymbol{D}(\boldsymbol{r},\, t) &= \rho, \\ \frac{\partial \boldsymbol{B}(\boldsymbol{r},t)}{\partial t} &= -\nabla \times \boldsymbol{E}(\boldsymbol{r},\, t), \\ \nabla \cdot \boldsymbol{B}(\boldsymbol{r},\, t) &= 0. \end{aligned} \tag{5}$$

Assuming the medium to be linear, we have

$$\begin{aligned} \boldsymbol{D}(\boldsymbol{r},\, t) &= \epsilon(\boldsymbol{r},\, t)\boldsymbol{E}(\boldsymbol{r},\, t), \\ \boldsymbol{B}(\boldsymbol{r},\, t) &= \mu(\boldsymbol{r},\, t)\boldsymbol{H}(\boldsymbol{r},\, t). \end{aligned} \tag{6}$$

Hereafter, we shall not indicate explicitly the space-time $(\boldsymbol{r},\, t)$ dependence of $\boldsymbol{E}$, $\boldsymbol{D}$, $\boldsymbol{B}$, $\boldsymbol{H}$, ϵ, μ, n, etc., unless required. Let us rewrite (5) as

$$\begin{aligned} \frac{\partial\left(\sqrt{\epsilon}\left(\sqrt{\epsilon}\,\boldsymbol{E}\right)\right)}{\partial t} &= \nabla \times \left(\frac{1}{\sqrt{\mu}}\left(\frac{\boldsymbol{B}}{\sqrt{\mu}}\right)\right) - \boldsymbol{J}, \\ \nabla \cdot \left(\sqrt{\mu}\left(\frac{\boldsymbol{B}}{\sqrt{\mu}}\right)\right) &= 0, \\ \frac{\partial}{\partial t}\left(\sqrt{\mu}\left(\frac{\boldsymbol{B}}{\sqrt{\mu}}\right)\right) &= -\nabla \times \left(\frac{1}{\sqrt{\epsilon}}\left(\sqrt{\epsilon}\,\boldsymbol{E}\right)\right), \\ \nabla \cdot \left(\sqrt{\epsilon}\left(\sqrt{\epsilon}\,\boldsymbol{E}\right)\right) &= \rho. \end{aligned} \tag{7}$$

For a medium with time-independent and spatially constant ϵ and μ (7) becomes

$$\begin{aligned} \frac{\partial\left(\sqrt{\epsilon}\,\boldsymbol{E}\right)}{\partial t} &= v \nabla \times \left(\frac{\boldsymbol{B}}{\sqrt{\mu}}\right) - \frac{\boldsymbol{J}}{\sqrt{\epsilon}}, \\ \nabla \cdot \left(\frac{\boldsymbol{B}}{\sqrt{\mu}}\right) &= 0, \\ \frac{\partial}{\partial t}\left(\frac{\boldsymbol{B}}{\sqrt{\mu}}\right) &= -v \nabla \times \left(\sqrt{\epsilon}\,\boldsymbol{E}\right), \\ \nabla \cdot \left(\sqrt{\epsilon}\,\boldsymbol{E}\right) &= \frac{\rho}{\sqrt{\epsilon}}, \end{aligned} \tag{8}$$

where $v = c/n$ is constant. Following (Bocker & Frieden, 1993, 2018), we write (8) in a matrix form as

$$\frac{\partial \mathcal{F}}{\partial t} = v \mathcal{M}_0 \mathcal{F} - \mathcal{J}, \tag{9a}$$

$$\mathcal{F} = \frac{1}{\sqrt{2}} \begin{pmatrix} \sqrt{\epsilon} E_x \\ \sqrt{\epsilon} E_y \\ \sqrt{\epsilon} E_z \\ 0 \\ \frac{1}{\sqrt{\mu}} B_x \\ \frac{1}{\sqrt{\mu}} B_y \\ \frac{1}{\sqrt{\mu}} B_z \\ 0 \end{pmatrix}, \qquad \mathcal{J} = \frac{1}{\sqrt{2\epsilon}} \begin{pmatrix} J_x \\ J_y \\ J_z \\ 0 \\ 0 \\ 0 \\ 0 \\ -v\rho \end{pmatrix}, \tag{9b}$$

$$\mathcal{M}_0 = \begin{pmatrix} 0 & 0 & 0 & 0 & 0 & -\partial_z & \partial_y & -\partial_x \\ 0 & 0 & 0 & 0 & \partial_z & 0 & -\partial_x & -\partial_y \\ 0 & 0 & 0 & 0 & -\partial_y & \partial_x & 0 & -\partial_z \\ 0 & 0 & 0 & 0 & \partial_x & \partial_y & \partial_z & 0 \\ 0 & \partial_z & -\partial_y & \partial_x & 0 & 0 & 0 & 0 \\ -\partial_z & 0 & \partial_x & \partial_y & 0 & 0 & 0 & 0 \\ \partial_y & -\partial_x & 0 & \partial_z & 0 & 0 & 0 & 0 \\ -\partial_x & -\partial_y & -\partial_z & 0 & 0 & 0 & 0 & 0 \end{pmatrix}, \tag{9c}$$

where $\partial_x = \partial/\partial x$, $\partial_y = \partial/\partial y$, and $\partial_z = \partial/\partial z$. Note that the elements of the matrix $\mathcal{M}_0$ are not fully determined by (8). The undetermined matrix elements, the elements of the 4th and 8th rows and columns, have been fixed by considerations of symmetry following (Bocker & Frieden, 1993) such that if we write

$$\mathcal{M}_0 = \mathcal{M}_{0x}\partial_x + \mathcal{M}_{0y}\partial_y + \mathcal{M}_{0z}\partial_z = \mathcal{M}_0 \cdot \nabla, \tag{10}$$

then the matrices $\mathcal{M}_{0x}$, $\mathcal{M}_{0y}$, and $\mathcal{M}_{0z}$ are Hermitian *i.e.*, $\mathcal{M}_{0x}^{\dagger} = \mathcal{M}_{0x}$, $\mathcal{M}_{0y}^{\dagger} = \mathcal{M}_{0y}$, $\mathcal{M}_{0z}^{\dagger} = \mathcal{M}_{0z}$. Explicitly writing, the matrices

$\mathcal{M}_{0x}$, $\mathcal{M}_{0y}$, and $\mathcal{M}_{0z}$ are:

$$\mathcal{M}_{0x} = \begin{pmatrix} 0 & 0 & 0 & 0 & 0 & 0 & 0 & -1 \\ 0 & 0 & 0 & 0 & 0 & 0 & -1 & 0 \\ 0 & 0 & 0 & 0 & 0 & 1 & 0 & 0 \\ 0 & 0 & 0 & 0 & 1 & 0 & 0 & 0 \\ 0 & 0 & 0 & 1 & 0 & 0 & 0 & 0 \\ 0 & 0 & 1 & 0 & 0 & 0 & 0 & 0 \\ 0 & -1 & 0 & 0 & 0 & 0 & 0 & 0 \\ -1 & 0 & 0 & 0 & 0 & 0 & 0 & 0 \end{pmatrix}, \tag{11a}$$

$$\mathcal{M}_{0y} = \begin{pmatrix} 0 & 0 & 0 & 0 & 0 & 0 & 1 & 0 \\ 0 & 0 & 0 & 0 & 0 & 0 & 0 & -1 \\ 0 & 0 & 0 & 0 & -1 & 0 & 0 & 0 \\ 0 & 0 & 0 & 0 & 0 & 1 & 0 & 0 \\ 0 & 0 & -1 & 0 & 0 & 0 & 0 & 0 \\ 0 & 0 & 0 & 1 & 0 & 0 & 0 & 0 \\ 1 & 0 & 0 & 0 & 0 & 0 & 0 & 0 \\ 0 & -1 & 0 & 0 & 0 & 0 & 0 & 0 \end{pmatrix}, \tag{11b}$$

$$\mathcal{M}_{0z} = \begin{pmatrix} 0 & 0 & 0 & 0 & 0 & -1 & 0 & 0 \\ 0 & 0 & 0 & 0 & 1 & 0 & 0 & 0 \\ 0 & 0 & 0 & 0 & 0 & 0 & 0 & -1 \\ 0 & 0 & 0 & 0 & 0 & 0 & 1 & 0 \\ 0 & 1 & 0 & 0 & 0 & 0 & 0 & 0 \\ -1 & 0 & 0 & 0 & 0 & 0 & 0 & 0 \\ 0 & 0 & 0 & 1 & 0 & 0 & 0 & 0 \\ 0 & 0 & -1 & 0 & 0 & 0 & 0 & 0 \end{pmatrix}. \tag{11c}$$

It is observed that

$$\begin{aligned} \mathcal{M}_{0x}^2 &= \mathcal{M}_{0y}^2 = \mathcal{M}_{0z}^2 = 1_8, \\ \mathcal{M}_{0x}\mathcal{M}_{0y} + \mathcal{M}_{0y}\mathcal{M}_{0x} &= 0_8, \\ \mathcal{M}_{0y}\mathcal{M}_{0z} + \mathcal{M}_{0z}\mathcal{M}_{0y} &= 0_8, \\ \mathcal{M}_{0z}\mathcal{M}_{0x} + \mathcal{M}_{0x}\mathcal{M}_{0z} &= 0_8, \end{aligned} \tag{12}$$

where 1_8 is the eight dimensional identity matrix and 0_8 is the eight dimensional null matrix. Then, it follows that

$$\mathcal{M}_0^2 = 1_8 \nabla^2. \tag{13}$$

In general, 1_n will denote an n-dimensional identity matrix and 0_n will denote an n-dimensional null matrix.

Let us now divide (9) throughout by v and rewrite it as

$$\left(\mathcal{M}_0 - \frac{1}{v}\frac{\partial}{\partial t}\right)\mathcal{F} = \frac{1}{v}\mathcal{J}. \tag{14}$$

Multiplying both sides of this equation by $(\mathcal{M}_0 + (1/v)\partial/\partial t)$ gives us

$$\left(\nabla^2 - \frac{1}{v^2}\frac{\partial^2}{\partial t^2}\right)\mathcal{F} = \frac{1}{v}\left(\mathcal{M}_0 + \frac{1}{v}\frac{\partial}{\partial t}\right)\mathcal{J}. \tag{15}$$

If we carry out the multiplication on the right hand side, then the inhomogeneous wave equations

$$\begin{aligned} \nabla^2\boldsymbol{E} - \frac{1}{v^2}\frac{\partial^2\boldsymbol{E}}{\partial t^2} &= \frac{1}{\epsilon_0}\boldsymbol{\nabla}\rho - \mu_0\frac{\partial\boldsymbol{J}}{\partial t}, \\ \nabla^2\boldsymbol{B} - \frac{1}{v^2}\frac{\partial^2\boldsymbol{B}}{\partial t^2} &= -\mu_0(\boldsymbol{\nabla}\times\boldsymbol{J}), \end{aligned} \tag{16}$$

and the charge continuity equation

$$\boldsymbol{\nabla}\cdot\boldsymbol{J} + \frac{\partial\rho}{\partial t} = 0, \tag{17}$$

follow. Thus, choosing the non-unique matrix elements of $\mathcal{M}_0$ such that $\mathcal{M}_{0x}$, $\mathcal{M}_{0y}$, and $\mathcal{M}_{0z}$ are Hermitian matrices makes (9) consistent with the inhomogeneous wave equations and the charge continuity equation. For a source-free medium (16) reduces to the wave equations

$$\begin{aligned} \nabla^2\boldsymbol{E} - \frac{1}{v^2}\frac{\partial^2\boldsymbol{E}}{\partial t^2} &= 0, \\ \nabla^2\boldsymbol{B} - \frac{1}{v^2}\frac{\partial^2\boldsymbol{B}}{\partial t^2} &= 0. \end{aligned} \tag{18}$$

In a homogeneous medium with $v = c/n$, where n is a constant, for a monochromatic light wave of circular frequency ω we have

$$\begin{aligned} \boldsymbol{E}(\boldsymbol{r}, t) &= \overline{\boldsymbol{E}}(\boldsymbol{r})e^{-i\omega t}, \\ \boldsymbol{B}(\boldsymbol{r}, t) &= \overline{\boldsymbol{B}}(\boldsymbol{r})e^{-i\omega t}. \end{aligned} \tag{19}$$

Then, the wave equations in (18) become, respectively,

$$\begin{aligned} (\nabla^2 + n^2k^2)\,\overline{\boldsymbol{E}} &= 0, \\ (\nabla^2 + n^2k^2)\,\overline{\boldsymbol{B}} &= 0, \end{aligned} \tag{20}$$

with $k = \omega/c = 2\pi/\lambda$, where k is the wave vector and λ is the wavelength of the monochromatic wave in vacuum. In the medium the circular frequency ω of the monochromatic wave is the same as when it is in vacuum and the wavelength and the wave vector become, respectively, $2\pi v/\omega = (2\pi c)/(n\omega) = \lambda/n$

and $\omega/v = (2\pi n)/\lambda = nk$. The time-independent wave equations for the monochromatic wave in (20) are the Helmholtz equations for a homogeneous medium of constant refractive index n. The Helmholtz equation

$$\left(\nabla^2 + n^2k^2\right)\Psi(\boldsymbol{r}) = 0, \tag{21}$$

is the basis of the scalar wave optics for a homogneous medium. The scalar function $\Psi(\boldsymbol{r})$ is taken to represent the complex amplitude of the monochromatic light wave at the point $\boldsymbol{r}$ with the intensity of light at that point given by $|\Psi(\boldsymbol{r})|^2$.

For a medium with ϵ and μ varying with space and time (7) becomes

$$\begin{aligned}
&\frac{\partial(\sqrt{\epsilon}\boldsymbol{E})}{\partial t} = -\frac{\dot{\epsilon}}{2\epsilon}(\sqrt{\epsilon}\boldsymbol{E}) + v\left(\boldsymbol{\nabla} - \frac{1}{2\mu}\boldsymbol{\nabla}\mu\right) \times \left(\frac{\boldsymbol{B}}{\sqrt{\mu}}\right) - \frac{\boldsymbol{J}}{\sqrt{\epsilon}},\\
&\left(\boldsymbol{\nabla} + \frac{1}{2\mu}\boldsymbol{\nabla}\mu\right)\cdot\left(\frac{\boldsymbol{B}}{\sqrt{\mu}}\right) = 0,\\
&\frac{\partial}{\partial t}\left(\frac{\boldsymbol{B}}{\sqrt{\mu}}\right) = -\frac{\dot{\mu}}{2\mu}\left(\frac{\boldsymbol{B}}{\sqrt{\mu}}\right) - v\left(\boldsymbol{\nabla} - \frac{1}{2\epsilon}\boldsymbol{\nabla}\epsilon\right) \times (\sqrt{\epsilon}\boldsymbol{E}),\\
&\left(\boldsymbol{\nabla} + \frac{1}{2\epsilon}\boldsymbol{\nabla}\epsilon\right)\cdot(\sqrt{\epsilon}\boldsymbol{E}) = \frac{\rho}{\sqrt{\epsilon}},
\end{aligned} \tag{22}$$

instead of (8). Defining

$$\bar{\epsilon} = \frac{1}{2}\ln\epsilon, \qquad \bar{\mu} = \frac{1}{2}\ln\mu, \tag{23}$$

(22) becomes, in terms of $\mathcal{F}$ and $\mathcal{J}$,

$$\frac{\partial\mathcal{F}}{\partial t} = \mathcal{M}\mathcal{F} - \mathcal{J}, \qquad \mathcal{M} = \begin{pmatrix} \mathcal{M}_{11} & \mathcal{M}_{12} \\ \mathcal{M}_{21} & \mathcal{M}_{22} \end{pmatrix}, \tag{24a}$$

$$\mathcal{M}_{11} = -\dot{\bar{\epsilon}}1_4, \qquad \mathcal{M}_{22} = -\dot{\bar{\mu}}1_4, \tag{24b}$$

$$\mathcal{M}_{12} = v\begin{pmatrix} 0 & -\partial_z + \partial_z\bar{\mu} & \partial_y - \partial_y\bar{\mu} & -\partial_x - \partial_x\bar{\mu} \\ \partial_z - \partial_z\bar{\mu} & 0 & -\partial_x + \partial_x\bar{\mu} & -\partial_y - \partial_y\bar{\mu} \\ -\partial_y + \partial_y\bar{\mu} & \partial_x - \partial_x\bar{\mu} & 0 & -\partial_z - \partial_z\bar{\mu} \\ \partial_x + \partial_x\bar{\mu} & \partial_y + \partial_y\bar{\mu} & \partial_z + \partial_z\bar{\mu} & 0 \end{pmatrix}, \tag{24c}$$

$$\mathcal{M}_{21} = v\begin{pmatrix} 0 & \partial_z - \partial_z\bar{\epsilon} & -\partial_y + \partial_y\bar{\epsilon} & \partial_x + \partial_x\bar{\epsilon} \\ -\partial_z + \partial_z\bar{\epsilon} & 0 & \partial_x - \partial_x\bar{\epsilon} & \partial_y + \partial_y\bar{\epsilon} \\ \partial_y - \partial_y\bar{\epsilon} & -\partial_x + \partial_x\bar{\epsilon} & 0 & \partial_z + \partial_z\bar{\epsilon} \\ -\partial_x - \partial_x\bar{\epsilon} & -\partial_y - \partial_y\bar{\epsilon} & -\partial_z - \partial_z\bar{\epsilon} & 0 \end{pmatrix}, \tag{24d}$$

with $\dot{\bar{\epsilon}} = \partial\bar{\epsilon}/\partial t$ and $\dot{\bar{\mu}} = \partial\bar{\mu}/\partial t$. Note that when ϵ and μ are constants $\mathcal{M} \longrightarrow \mathcal{M}_0$.

When ϵ, μ, and hence v, are time-independent and the medium is source-free we have

$$\frac{\partial \mathcal{F}}{\partial t} = \mathcal{M}\mathcal{F} = \begin{pmatrix} \mathbf{0} & \mathcal{M}_{12} \\ \mathcal{M}_{21} & \mathbf{0} \end{pmatrix} \mathcal{F}. \tag{25}$$

For such a medium with time-independent $\mathcal{M}$ we can write

$$\begin{aligned} \frac{\partial^2 \mathcal{F}}{\partial t^2} &= \begin{pmatrix} \mathbf{0} & \mathcal{M}_{12} \\ \mathcal{M}_{21} & \mathbf{0} \end{pmatrix} \begin{pmatrix} \mathbf{0} & \mathcal{M}_{12} \\ \mathcal{M}_{21} & \mathbf{0} \end{pmatrix} \mathcal{F} \\ &= \begin{pmatrix} \mathcal{M}_{12}\mathcal{M}_{21} & \mathbf{0} \\ \mathbf{0} & \mathcal{M}_{21}\mathcal{M}_{12} \end{pmatrix} \mathcal{F}. \end{aligned} \tag{26}$$

As shown in Appendix A, (26) leads to the wave equation for $\boldsymbol{E}$ in an inhomogeneous medium (Born & Wolf, 1999):

$$\nabla^2 \boldsymbol{E} - \epsilon\mu \frac{\partial^2 \boldsymbol{E}}{\partial t^2} = (\nabla \times \boldsymbol{E}) \times (\nabla \ln \mu) - \nabla\left((\nabla \ln \epsilon)\cdot \boldsymbol{E}\right). \tag{27}$$

Similar equation can be obtained for the magnetic field. When the ϵ and μ are slowly varying functions of $\boldsymbol{r}$, such that $\nabla \ln \mu| \ll 1$ and $|\nabla \ln \mu| \ll 1$, we can approximate (27) as

$$\nabla^2 \boldsymbol{E} - \epsilon\mu \frac{\partial^2 \boldsymbol{E}}{\partial t^2} = 0, \tag{28}$$

or

$$\nabla^2 \boldsymbol{E} - \frac{n^2(\boldsymbol{r})}{c^2} \frac{\partial^2 \boldsymbol{E}}{\partial t^2} = 0. \tag{29}$$

For a monochromatic wave with $\boldsymbol{E}(\boldsymbol{r}, t) = \overline{\boldsymbol{E}}(\boldsymbol{r}) e^{-i\omega t}$, we get

$$\left(\nabla^2 + n^2(\boldsymbol{r}) k^2\right) \overline{\boldsymbol{E}} = 0, \tag{30}$$

the Helmholtz equation for a source-free inhomogeneous medium, where $k = \omega/c$ is the wave vector in vacuum. In the scalar wave optics, for a time-independent inhomogeneous medium of refractive index $n(\boldsymbol{r})$ one takes

$$\left(\nabla^2 + n^2(\boldsymbol{r}) k^2\right) \Psi(\boldsymbol{r}) = 0, \tag{31}$$

where the complex function $\Psi(\boldsymbol{r})$ is the amplitude, and $|\Psi(\boldsymbol{r})|^2$ is the intensity, of the monochromatic light wave moving in the medium. Note that for a homogeneous medium $n(\boldsymbol{r})$ becomes a constant and (31) reduces to (21).

Let us now define

$$\underline{\Psi} = \mathbb{T}\mathcal{F} = \frac{1}{2}\begin{pmatrix} -F_x^+ + iF_y^+ \\ F_z^+ \\ F_z^+ \\ F_x^+ + iF_y^+ \\ -F_x^- - iF_y^- \\ F_z^- \\ F_z^- \\ F_x^- - iF_y^- \end{pmatrix}, \qquad \mathfrak{J} = \mathbb{T}\mathcal{J} = \frac{1}{2\sqrt{2\epsilon}}\begin{pmatrix} -J_x + iJ_y \\ J_z + v\rho \\ J_z - v\rho \\ J_x + iJ_y \\ -J_x - iJ_y \\ J_z + v\rho \\ J_z - v\rho \\ J_x - iJ_y \end{pmatrix}, \tag{32}$$

where

$$\boldsymbol{F}^{\pm} = \frac{1}{\sqrt{2}}\left(\sqrt{\epsilon}\,\boldsymbol{E} \pm \frac{i}{\sqrt{\mu}}\boldsymbol{B}\right) = \sqrt{\frac{\epsilon}{2}}\,(\boldsymbol{E} \pm iv\boldsymbol{B}), \tag{33}$$

are the Riemann-Silberstein-Weber (RSW) vectors, and

$$\mathbb{T} = \frac{1}{2}\begin{pmatrix} -1 & i & 0 & 0 & -i & -1 & 0 & 0 \\ 0 & 0 & 1 & i & 0 & 0 & i & -1 \\ 0 & 0 & 1 & -i & 0 & 0 & i & 1 \\ 1 & i & 0 & 0 & i & -1 & 0 & 0 \\ -1 & -i & 0 & 0 & i & -1 & 0 & 0 \\ 0 & 0 & 1 & -i & 0 & 0 & -i & -1 \\ 0 & 0 & 1 & i & 0 & 0 & -i & 1 \\ 1 & -i & 0 & 0 & -i & -1 & 0 & 0 \end{pmatrix}. \tag{34}$$

In Appendix B we list the properties of the RSW vectors in some detail. For a detailed discussion of the Riemann-Silberstein-Weber complex vector, see (Bialynicki-Birula, 1994, 1996a, 1996b). Note that $\mathbb{T}$ is a unitary matrix *i.e.*, $\mathbb{T}^{\dagger}\mathbb{T} = \mathbb{T}\mathbb{T}^{\dagger} = 1_8$.

Let us recall that the three Pauli matrices are given by

$$\sigma_x = \begin{pmatrix} 0 & 1 \\ 1 & 0 \end{pmatrix}, \quad \sigma_y = \begin{pmatrix} 0 & -i \\ i & 0 \end{pmatrix}, \quad \sigma_z = \begin{pmatrix} 1 & 0 \\ 0 & -1 \end{pmatrix}, \tag{35}$$

which are Hermitian *i.e.*, $\sigma_j^\dagger = \sigma_j$, $j = x, y, z$. The Pauli matrices obey the algebra

$$\begin{aligned} \sigma_x^2 = \sigma_y^2 = \sigma_z^2 = 1_2, \\ \sigma_x\sigma_y + \sigma_y\sigma_x = 0_2, \\ \sigma_y\sigma_z + \sigma_z\sigma_y = 0_2, \\ \sigma_z\sigma_x + \sigma_x\sigma_z = 0_2. \end{aligned} \tag{36}$$

Note that the algebra (12) obeyed by the matrices $\{\mathcal{M}_{0x}, \mathcal{M}_{0y}, \mathcal{M}_{0z}\}$ is the same as the Pauli algebra (36). We shall write

$$\begin{aligned} \sigma_x\partial_x + \sigma_y\partial_y + \sigma_z\partial_z &= \boldsymbol{\sigma}\cdot\nabla, \\ \sigma_x^*\partial_x + \sigma_y^*\partial_y + \sigma_z^*\partial_z &= \boldsymbol{\sigma}^*\cdot\nabla, \end{aligned} \tag{37}$$

where * denotes the complex conjugate, and $\sigma_x^* = \sigma_x$, $\sigma_y^* = -\sigma_y$, $\sigma_z^* = \sigma_z$. Writing (24) in terms of $\underline{\Psi}$ and $\mathfrak{J}$ we get

$$\frac{\partial\underline{\Psi}}{\partial t} = (\mathbb{T}\mathcal{M}\mathbb{T}^\dagger)\underline{\Psi} - \mathfrak{J} = \mathsf{M}\underline{\Psi} - \mathfrak{J}, \quad \mathsf{M} = v\mathsf{M}_0 + \mathsf{M}', \tag{38a}$$

$$\mathsf{M}_0 = -\begin{pmatrix} \boldsymbol{\sigma}\cdot\nabla & 0_2 & 0_2 & 0_2 \\ 0_2 & \boldsymbol{\sigma}\cdot\nabla & 0_2 & 0_2 \\ 0_2 & 0_2 & \boldsymbol{\sigma}^*\cdot\nabla & 0_2 \\ 0_2 & 0_2 & 0_2 & \boldsymbol{\sigma}^*\cdot\nabla \end{pmatrix}, \tag{38b}$$

$$\mathsf{M}' = \begin{pmatrix} \mathsf{M}'_{11} & \mathsf{M}'_{12} \\ \mathsf{M}'_{21} & \mathsf{M}'_{22} \end{pmatrix} \tag{38c}$$

$$\mathsf{M}'_{11} = \frac{v}{2}\begin{pmatrix} \partial_z(\bar{\epsilon} + \bar{\mu})1_2 & \partial_-(\bar{\epsilon} + \bar{\mu})1_2 \\ \partial_+(\bar{\epsilon} + \bar{\mu})1_2 & -\partial_z(\bar{\epsilon} + \bar{\mu})1_2 \end{pmatrix} - \frac{1}{2}(\dot{\bar{\epsilon}} + \dot{\bar{\mu}})1_4 \tag{38d}$$

$$\mathsf{M}'_{22} = \frac{v}{2}\begin{pmatrix} \partial_z(\bar{\epsilon} + \bar{\mu})1_2 & \partial_+(\bar{\epsilon} + \bar{\mu})1_2 \\ \partial_-(\bar{\epsilon} + \bar{\mu})1_2 & -\partial_z(\bar{\epsilon} + \bar{\mu})1_2 \end{pmatrix} - \frac{1}{2}(\dot{\bar{\epsilon}} + \dot{\bar{\mu}})1_4 \tag{38e}$$

$$\begin{aligned} \mathsf{M}'_{12} = &\frac{v}{2}\begin{pmatrix} i\partial_-(\bar{\epsilon} - \bar{\mu})\sigma_y & -i\partial_z(\bar{\epsilon} - \bar{\mu})\sigma_y \\ -i\partial_z(\bar{\epsilon} - \bar{\mu})\sigma_y & -i\partial_+(\bar{\epsilon} - \bar{\mu})\sigma_y \end{pmatrix} \\ &-\frac{1}{2}(\dot{\bar{\epsilon}} - \dot{\bar{\mu}})\begin{pmatrix} 0_2 & -i\sigma_y \\ i\sigma_y & 0_2 \end{pmatrix}, \end{aligned} \tag{38f}$$

$$\begin{aligned}\mathsf{M}'_{21} &= \frac{v}{2}\begin{pmatrix} i\partial_+(\bar{\epsilon} - \bar{\mu})\sigma_y & -i\partial_z(\bar{\epsilon} - \bar{\mu})\sigma_y \\ -i\partial_z(\bar{\epsilon} - \bar{\mu})\sigma_y & -i\partial_-(\bar{\epsilon} - \bar{\mu})\sigma_y \end{pmatrix} \\ &\quad -\frac{1}{2}(\dot{\bar{\epsilon}} - \dot{\bar{\mu}})\begin{pmatrix} 0_2 & -i\sigma_y \\ i\sigma_y & 0_2 \end{pmatrix}.\end{aligned} \tag{38g}$$

Let us now write (38) in terms of the refractive index n and the impedance η of the medium. To this end, we define

$$\bar{n} = \frac{1}{2}\ln n, \qquad \bar{\eta} = \frac{1}{2}\ln \eta, \tag{39}$$

and use the relations

$$\epsilon = \frac{n}{c\eta}, \qquad \mu = \frac{n\eta}{c}, \qquad v = \frac{c}{n}. \tag{40}$$

Then, we have

$$\begin{aligned}\dot{\bar{\epsilon}} + \dot{\bar{\mu}} &= 2\,\dot{\bar{n}}, \qquad \dot{\bar{\epsilon}} - \dot{\bar{\mu}} = -2\,\dot{\bar{\eta}}, \\ \nabla(\bar{\epsilon} + \bar{\mu}) &= 2\,\nabla\bar{n}, \qquad \nabla(\bar{\epsilon} - \bar{\mu}) = -2\,\nabla\bar{\eta}.\end{aligned} \tag{41}$$

Using these relations, (38) becomes

$$\frac{\partial \underline{\Psi}}{\partial t} = M\underline{\Psi} - \mathfrak{J}, \qquad M = vM_0 + M', \tag{42a}$$

$$M_0 = -\begin{pmatrix} \boldsymbol{\sigma}\cdot\nabla & 0_2 & 0_2 & 0_2 \\ 0_2 & \boldsymbol{\sigma}\cdot\nabla & 0_2 & 0_2 \\ 0_2 & 0_2 & \boldsymbol{\sigma}^*\cdot\nabla & 0_2 \\ 0_2 & 0_2 & 0_2 & \boldsymbol{\sigma}^*\cdot\nabla \end{pmatrix}, \tag{42b}$$

$$\begin{aligned}M' &= \frac{c}{n}\begin{pmatrix} (\partial_z\bar{n})\,1_2 & (\partial_-\bar{n})\,1_2 & -i(\partial_-\bar{\eta})\sigma_y & i(\partial_z\bar{\eta})\sigma_y \\ (\partial_+\bar{n})\,1_2 & -(\partial_z\bar{n})\,1_2 & i(\partial_z\bar{\eta})\sigma_y & i(\partial_+\bar{\eta})\sigma_y \\ -i(\partial_+\bar{\eta})\sigma_y & i(\partial_z\bar{\eta})\sigma_y & (\partial_z\bar{n})\,1_2 & (\partial_+\bar{n})\,1_2 \\ i(\partial_z\bar{\eta})\sigma_y & i(\partial_-\bar{\eta})\sigma_y & (\partial_-\bar{n})\,1_2 & -(\partial_z\bar{n})\,1_2 \end{pmatrix} \\ &\quad -\frac{\dot{n}}{2n}1_8 + \frac{\dot{\eta}}{2\eta}\begin{pmatrix} 0_2 & 0_2 & 0_2 & -i\sigma_y \\ 0_2 & 0_2 & i\sigma_y & 0_2 \\ 0_2 & -i\sigma_y & 0_2 & 0_2 \\ i\sigma_y & 0_2 & 0_2 & 0_2 \end{pmatrix}.\end{aligned} \tag{42c}$$

For a source-free medium with time-independent n and η the time evolution equation for $\underline{\Psi}$ becomes

$$\frac{\partial \underline{\Psi}}{\partial t} = M\underline{\Psi}, \qquad M = M_0 + M', \tag{43a}$$

$$M_0 = -v\begin{pmatrix} \boldsymbol{\sigma}\cdot\boldsymbol{\nabla} & 0_2 & 0_2 & 0_2 \\ 0_2 & \boldsymbol{\sigma}\cdot\boldsymbol{\nabla} & 0_2 & 0_2 \\ 0_2 & 0_2 & \boldsymbol{\sigma}^*\cdot\boldsymbol{\nabla} & 0_2 \\ 0_2 & 0_2 & 0_2 & \boldsymbol{\sigma}^*\cdot\boldsymbol{\nabla} \end{pmatrix}, \tag{43b}$$

$$M' = \frac{c}{n}\begin{pmatrix} (\partial_z\bar{n})\,1\!\!1_2 & (\partial_-\bar{n})\,1\!\!1_2 & -i\,(\partial_-\bar{\eta})\sigma_y & i\,(\partial_z\bar{\eta})\sigma_y \\ (\partial_+\bar{n})\,1\!\!1_2 & -(\partial_z\bar{n})\,1\!\!1_2 & i\,(\partial_z\bar{\eta})\sigma_y & i\,(\partial_+\bar{\eta})\sigma_y \\ -i\,(\partial_+\bar{\eta})\sigma_y & i\,(\partial_z\bar{\eta})\sigma_y & (\partial_z\bar{n})\,1\!\!1_2 & (\partial_+\bar{n})\,1\!\!1_2 \\ i\,(\partial_z\bar{\eta})\sigma_y & i\,(\partial_-\bar{\eta})\sigma_y & (\partial_-\bar{n})\,1\!\!1_2 & -(\partial_z\bar{n})\,1\!\!1_2 \end{pmatrix}. \tag{43c}$$

Equation (38), or (42), is, in general, completely equivalent to the Maxwell equations (5). When ϵ and μ are time-independent and spatially constant in a medium, M' vanishes and we have

$$\begin{aligned} \frac{\partial \underline{\Psi}}{\partial t} &= vM_0\underline{\Psi} - \mathfrak{J}, \\ M_0 &= -\begin{pmatrix} \boldsymbol{\sigma}\cdot\boldsymbol{\nabla} & 0_2 & 0_2 & 0_2 \\ 0_2 & \boldsymbol{\sigma}\cdot\boldsymbol{\nabla} & 0_2 & 0_2 \\ 0_2 & 0_2 & \boldsymbol{\sigma}^*\cdot\boldsymbol{\nabla} & 0_2 \\ 0_2 & 0_2 & 0_2 & \boldsymbol{\sigma}^*\cdot\boldsymbol{\nabla} \end{pmatrix}, \end{aligned} \tag{44}$$

equivalent to the Maxwell equations corresponding to a homogneous medium. Thus, to reduce $\mathcal{M}_0$ in (9) to the simple block diagonal form M_0 in (44) the change of basis of the matrix representation of the Maxwell equations from the column vector $\mathcal{F}$ based on the canonical field vectors to the column vector $\underline{\Psi}$ based on the RSW-vectors is seen to be the natural choice. A complete derivation of this reduction based on the unitary transformation matrix $\mathbb{T}$ is given in (Khan & Jagannathan, 2024c).

In order to make the matrix representation of the Maxwell equations in (38) better adapted to analyse beam propagation in the framework of the

Maxwell vector wave optics we shall now introduce a further transformation of $\underline{\Psi}$. Let

$$\Phi(\boldsymbol{r}, t) = \mathcal{S}\underline{\Psi}(\boldsymbol{r}, t),$$

$$\mathcal{S} = \begin{pmatrix} 1 & 0 & 0 & 0 & 0 & 0 & 0 & 0 \\ 0 & 0 & 1 & 0 & 0 & 0 & 0 & 0 \\ 0 & 0 & 0 & 0 & 1 & 0 & 0 & 0 \\ 0 & 0 & 0 & 0 & 0 & 0 & 1 & 0 \\ 0 & 1 & 0 & 0 & 0 & 0 & 0 & 0 \\ 0 & 0 & 0 & 1 & 0 & 0 & 0 & 0 \\ 0 & 0 & 0 & 0 & 0 & 1 & 0 & 0 \\ 0 & 0 & 0 & 0 & 0 & 0 & 0 & 1 \end{pmatrix}. \tag{45}$$

This makes

$$\Phi = \begin{pmatrix} \phi^+ \\ \phi^- \end{pmatrix},$$

$$\phi^+ = \frac{1}{2}\begin{pmatrix} -F_x^+ + iF_y^+ \\ F_z^+ \\ -F_x^- - iF_y^- \\ F_z^- \end{pmatrix}, \quad \phi^- = \frac{1}{2}\begin{pmatrix} F_z^+ \\ F_x^+ + iF_y^+ \\ F_z^- \\ F_x^- - iF_y^- \end{pmatrix}. \tag{46}$$

The fields $\boldsymbol{E}$ and $\boldsymbol{B}$ can be recovered from Φ using the inverse transformation:

$$\mathcal{F} = \mathbb{T}^\dagger \mathcal{S}^\dagger \Phi = \mathcal{T}\Phi,$$

$$\mathcal{T} = \begin{pmatrix} -1 & 0 & -1 & 0 & 0 & 1 & 0 & 1 \\ -i & 0 & i & 0 & 0 & -i & 0 & i \\ 0 & 1 & 0 & 1 & 1 & 0 & 1 & 0 \\ 0 & i & 0 & -i & -i & 0 & i & 0 \\ i & 0 & -i & 0 & 0 & -i & 0 & i \\ -1 & 0 & -1 & 0 & 0 & -1 & 0 & -1 \\ 0 & -i & 0 & i & -i & 0 & i & 0 \\ 0 & 1 & 0 & 1 & -1 & 0 & -1 & 0 \end{pmatrix} \tag{47}$$

Note that $\mathcal{T}$ is unitary: $\mathcal{T}^\dagger\mathcal{T} = \mathcal{T}\mathcal{T}^\dagger = 1_8$. From now on, we shall consider only a medium which is source-free and has time-independent n and η. Then, the time evolution equation satisfied by Φ, as obtained from (43), is

$$\frac{\partial \Phi}{\partial t} = (\mathcal{S}M\mathcal{S}^\dagger)\Phi = \mathbb{M}\Phi, \quad \mathbb{M} = \mathbb{M}_0 + \mathbb{M}', \tag{48a}$$

$$\mathbb{M}_0 = -\frac{c}{n}\begin{pmatrix} 1_2\partial_z & 0_2 & 1_2\partial_- & 0_2 \\ 0_2 & 1_2\partial_z & 0_2 & 1_2\partial_+ \\ 1_2\partial_+ & 0_2 & -1_2\partial_z & 0_2 \\ 0_2 & 1_2\partial_- & 0_2 & -1_2\partial_z \end{pmatrix}, \tag{48b}$$

$$\mathbb{M}' = \frac{c}{n}\begin{pmatrix} \boldsymbol{\sigma}\cdot\nabla\bar{n} & 0_2 & 0_2 & -i\sigma_y(\boldsymbol{\sigma}^*\cdot\nabla\bar{\eta}) \\ 0_2 & \boldsymbol{\sigma}^*\cdot\nabla\bar{n} & -i\sigma_y(\boldsymbol{\sigma}\cdot\nabla\bar{\eta}) & 0_2 \\ 0_2 & i\sigma_y(\boldsymbol{\sigma}^*\cdot\nabla\bar{\eta}) & \boldsymbol{\sigma}\cdot\nabla\bar{n} & 0_2 \\ i\sigma_y(\boldsymbol{\sigma}\cdot\nabla\bar{\eta}) & 0_2 & 0_2 & \boldsymbol{\sigma}^*\cdot\nabla\bar{n} \end{pmatrix}. \tag{48c}$$

Let us now consider a paraxial monochromatic beam of circular frequency ω propagating predominantly along the z-direction in an inhomogeneous medium with refractive index $n(\boldsymbol{r}_\perp; z)$. We can take the field associated with the beam to be

$$\begin{aligned} \boldsymbol{E}(\boldsymbol{r}, t) &= \overline{\boldsymbol{E}}(\boldsymbol{r}_\perp; z)\, e^{-i\omega t}, \\ \boldsymbol{B}(\boldsymbol{r}, t) &= \overline{\boldsymbol{B}}(\boldsymbol{r}_\perp; z)\, e^{-i\omega t}, \end{aligned} \tag{49}$$

where $\boldsymbol{r}_\perp = (x, y)$, and $(\overline{\boldsymbol{E}}(\boldsymbol{r}_\perp; z), \overline{\boldsymbol{B}}(\boldsymbol{r}_\perp; z))$ is the field in the xy-plane at the point z on the axis of the beam at $t = 0$. Correspondingly, we can write

$$\boldsymbol{\Phi}(\boldsymbol{r}, t) = \begin{pmatrix} \overline{\phi}^+(\boldsymbol{r}_\perp; z) \\ \overline{\phi}^-(\boldsymbol{r}_\perp; z) \end{pmatrix} e^{-i\omega t} = \overline{\boldsymbol{\Phi}}(\boldsymbol{r}_\perp; z)\, e^{-i\omega t}. \tag{50}$$

Substituting this $\boldsymbol{\Phi}(\boldsymbol{r}, t)$ in (48) and rearranging the resulting terms we get

$$\begin{aligned} &\frac{c}{n}\begin{pmatrix} 1_4\partial_z & 0_4 \\ 0_4 & -1_4\partial_z \end{pmatrix}\begin{pmatrix} \overline{\phi}^+(\boldsymbol{r}_\perp; z) \\ \overline{\phi}^-(\boldsymbol{r}_\perp; z) \end{pmatrix} \\ &= \left[i\omega 1_8 - \frac{c}{n}\begin{pmatrix} 0_2 & 0_2 & 1_2\partial_- & 0_2 \\ 0_2 & 0_2 & 0_2 & 1_2\partial_+ \\ 1_2\partial_+ & 0_2 & 0_2 & 0_2 \\ 0_2 & 1_2\partial_- & 0_2 & 0_2 \end{pmatrix} + \mathbb{M}' \right] \\ &\quad \begin{pmatrix} \overline{\phi}^+(\boldsymbol{r}_\perp; z) \\ \overline{\phi}^-(\boldsymbol{r}_\perp; z) \end{pmatrix}. \end{aligned} \tag{51}$$

Let us define the eight dimensional analog of Dirac's β matrix as

$$\mathcal{B} = \begin{pmatrix} 1_4 & 0_4 \\ 0_4 & -1_4 \end{pmatrix}. \tag{52}$$

Dirac's β matrix is

$$\beta = \begin{pmatrix} 1_2 & 0_2 \\ 0_2 & -1_2 \end{pmatrix}. \tag{53}$$

For the monochromatic beam $\lambda = 2\pi c/\omega$ is the wavelength while traveling in vacuum and $k = 2\pi/\lambda = \omega/c$ is the magnitude of the corresponding wave vector. If we write $\lambda/2\pi = \lambdabar$, then $c/\omega = \lambdabar = 1/k$. Multiplying both sides of (51) from left by $i(n/\omega)\mathcal{B}$ and rearranging the terms we get the Maxwell vector wave optical z-evolution equation

$$i\lambdabar \frac{\partial \overline{\Phi}(\boldsymbol{r}_\perp; z)}{\partial z} = \widehat{\mathcal{H}}(\boldsymbol{r}_\perp, \widehat{\wp}_\perp, z)\overline{\Phi}(\boldsymbol{r}_\perp; z), \tag{54a}$$

$$\widehat{\mathcal{H}} = -n_c\mathcal{B} + \widehat{\mathcal{E}} + \widehat{\mathcal{O}}, \tag{54b}$$

$$\widehat{\mathcal{E}} = -(\delta n)\,\mathcal{B} + i\lambdabar \begin{pmatrix} \boldsymbol{\sigma}\cdot\nabla\overline{n} & 0_2 & 0_2 & 0_2 \\ 0_2 & \boldsymbol{\sigma}^*\cdot\nabla\overline{n} & 0_2 & 0_2 \\ 0_2 & 0_2 & -\boldsymbol{\sigma}\cdot\nabla\overline{n} & 0_2 \\ 0_2 & 0_2 & 0_2 & -\boldsymbol{\sigma}^*\cdot\nabla\overline{n} \end{pmatrix}, \tag{54c}$$

$$\widehat{\mathcal{O}} = \begin{pmatrix} 0_2 & 0_2 & 1_2\widehat{\wp}_- & \lambdabar\sigma_y \times (\boldsymbol{\sigma}^*\cdot\nabla\overline{\eta}) \\ 0_2 & 0_2 & \lambdabar\sigma_y \times (\boldsymbol{\sigma}\cdot\nabla\overline{\eta}) & 1_2\widehat{\wp}_+ \\ -1_2\widehat{\wp}_+ & \lambdabar\sigma_y \times (\boldsymbol{\sigma}^*\cdot\nabla\overline{\eta}) & 0_2 & 0_2 \\ \lambdabar\sigma_y \times (\boldsymbol{\sigma}\cdot\nabla\overline{\eta}) & -1_2\widehat{\wp}_- & 0_2 & 0_2 \end{pmatrix}, \tag{54d}$$

where we have written $n(\boldsymbol{r}_\perp; z) = n_c + \delta n(\boldsymbol{r}_\perp; z)$, assuming that the refractive index fluctuates around a constant value n_c, and

$$\begin{aligned} \widehat{\wp}_\perp &= (\widehat{\wp}_x, \widehat{\wp}_y), \\ \widehat{\wp}_x &= -i\lambdabar\frac{\partial}{\partial x}, \quad \widehat{\wp}_y = -i\lambdabar\frac{\partial}{\partial y}, \\ \widehat{\wp}_+ &= \widehat{\wp}_x + i\widehat{\wp}_y, \quad \widehat{\wp}_- = \widehat{\wp}_x - i\widehat{\wp}_y. \end{aligned} \tag{55}$$

Note that

$$\mathcal{B}\widehat{\mathcal{E}} = \widehat{\mathcal{E}}\mathcal{B}, \qquad \mathcal{B}\widehat{\mathcal{O}} = -\widehat{\mathcal{O}}\mathcal{B}. \tag{56}$$

The Maxwell vector wave optical Hamiltonian $\widehat{\mathcal{H}}(\boldsymbol{r}_\perp, \widehat{\wp}_\perp, z)$, generates the z-evolution of $\overline{\boldsymbol{\Phi}}(\boldsymbol{r}_\perp; z)$ representing the beam field. Formally integrating (54a), we have

$$\overline{\boldsymbol{\Phi}}(\boldsymbol{r}_\perp; z) = \widehat{\mathcal{T}}(z, z')\overline{\boldsymbol{\Phi}}(\boldsymbol{r}_\perp; z') \tag{57a}$$

$$i\lambda\!\!\!{}^{-}\frac{\partial}{\partial z}\widehat{\mathcal{T}}(z, z') = \widehat{\mathcal{H}}\,\widehat{\mathcal{T}}(z, z'), \quad \widehat{\mathcal{T}}(z', z') = I, \tag{57b}$$

$$\begin{aligned} \widehat{\mathcal{T}}(z, z') &= I - \tfrac{i}{\lambda\!\!\!{}^{-}}\int_{z'}^{z} dz_1\,\widehat{\mathcal{H}}(z_1) \\ &\quad + \left(-\tfrac{i}{\lambda\!\!\!{}^{-}}\right)^2 \int_{z'}^{z} dz_2 \int_{z'}^{z_2} dz_1\,\widehat{\mathcal{H}}(z_2)\widehat{\mathcal{H}}(z_1) + \cdots \\ &= \mathbb{P}\left[\exp\left(-\tfrac{i}{\lambda\!\!\!{}^{-}}\int_{z'}^{z} dz\,\widehat{\mathcal{H}}(z)\right)\right], \end{aligned} \tag{57c}$$

where I is the identity operator and $\mathbb{P}$ denotes the path-ordered exponential. From

$$\begin{aligned} & i\lambda\!\!\!{}^{-}\tfrac{\partial}{\partial z}\left[\left(\widehat{\mathcal{T}}(z, z')\right)^{-1}\widehat{\mathcal{T}}(z, z')\right] \\ &= i\lambda\!\!\!{}^{-}\tfrac{\partial}{\partial z}\left[\left(\widehat{\mathcal{T}}(z, z')\right)^{-1}\right]\widehat{\mathcal{T}}(z, z') + \left(\widehat{\mathcal{T}}(z, z')\right)^{-1}\widehat{\mathcal{H}}\widehat{\mathcal{T}}(z, z') \\ &= 0, \end{aligned} \tag{58}$$

it follows that

$$i\lambda\!\!\!{}^{-}\frac{\partial}{\partial z}\left(\left(\widehat{\mathcal{T}}(z, z')\right)^{-1}\right) = -\left(\widehat{\mathcal{T}}(z, z')\right)^{-1}\widehat{\mathcal{H}}. \tag{59}$$

In general, $\widehat{\mathcal{T}}(z, z')$ does not have a closed form expression. The most useful expression for the z-evolution operator $\widehat{\mathcal{T}}(z, z')$, or the z-propagator, is its exponential form:

$$\widehat{\mathcal{T}}(z, z') = \exp\left(-\frac{i}{\lambda\!\!\!{}^{-}}\widehat{T}(z, z')\right), \tag{60a}$$

$$\begin{aligned} \widehat{T}(z, z') &= \int_{z'}^{z} dz_1\,\widehat{\mathcal{H}}(z_1) \\ &\quad + \tfrac{1}{2}\left(-\tfrac{i}{\lambda\!\!\!{}^{-}}\right)\int_{z'}^{z} dz_2 \int_{z'}^{z_2} dz_1\left[\widehat{\mathcal{H}}(z_2), \widehat{\mathcal{H}}(z_1)\right] + \ldots, \end{aligned} \tag{60b}$$

where $[\widehat{A}, \widehat{B}] = \widehat{A}\widehat{B} - \widehat{B}\widehat{A}$, the commutator of $\widehat{A}$ and $\widehat{B}$. The series expression in (60) is the Magnus formula (Blanes et al., 2009; Magnus,

1954; Mananga & Charpentier, 2016; Wilcox, 1967). The Magnus formula is required for studying propagation in media with spatially varying refractive index.

We can write (57), the z-evolution equation for $\overline{\Phi}(\boldsymbol{r}_\perp; z)$, in an integral form:

$$\overline{\Phi}(\boldsymbol{r}_\perp; z) = \int d^2r'_\perp \, K(\boldsymbol{r}_\perp, z; \boldsymbol{r}'_\perp, z')\overline{\Phi}(\boldsymbol{r}'_\perp; z'), \tag{61}$$

where $\int d^2r'_\perp$ stands for $\int\int dx'dy'$ with integration over the entire $x'y'$ plane, and $K(\boldsymbol{r}_\perp, z; \boldsymbol{r}'_\perp, z')$ is the Point Spread Function (PSF) of the system. It should be noted that, since $\overline{\Phi}$ is an eight dimensional column vector function, $K(\boldsymbol{r}_\perp, z; \boldsymbol{r}'_\perp, z')$ is an eight dimensional matrix function. This generalizes the PSF of the Helmholtz scalar wave optics to the Maxwell vector wave optics. Equation (61) represents the Fresnel-Kirchhoff diffraction formula in the Maxwell vector wave optics (see, e.g., Born & Wolf, 1999). In Section 5 we will discuss the PSF for the propagation of a paraxial beam in a medium with constant refractive index.

Thus, to study the z-propagation of a monochromatic electromagnetic beam represented by the canonical field vector $\mathcal{F}$ from the point z' to the point z we have to use the following scheme:

$$\begin{aligned} \mathcal{F}(z') &\longrightarrow \mathcal{T}^\dagger\mathcal{F}(z') = \overline{\Phi}(z')e^{-i\omega t}, \\ \overline{\Phi}(z') &\longrightarrow \overline{\Phi}(z) = \widehat{\mathcal{T}}(z, z')\overline{\Phi}(z'), \\ e^{-i\omega t}\overline{\Phi}(z) &\longrightarrow e^{-i\omega t}\mathcal{T}\overline{\Phi}(z) = \mathcal{F}(z). \end{aligned} \tag{62}$$

To study the propagation of a general electromagnetic wave one can use its decomposition into monochromatic plane waves.

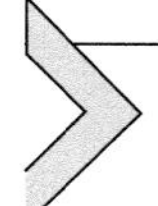

3. Foldy-Wouthuysen-like successive approximation technique for beam optical systems

It is seen that the Maxwell vector wave optical Hamiltonian $\widehat{\mathcal{H}}$ in (54) has the structure of the Dirac Hamiltonian for the electron (see, e.g., Bjorken & Drell, 1994; Greiner, 2000). Apart from the constant term $-n_c\mathcal{B}$, $\widehat{\mathcal{H}}$ is the sum of an 'even' operator $\widehat{\mathcal{E}}$, which does not couple the upper four components and the lower four components of $\overline{\Phi}$, and an 'odd' operator $\widehat{\mathcal{O}}$, which couples the upper four components and the lower four components of $\overline{\Phi}$. Then, it becomes clear that to study the propagation of the beam along the z-axis one can expand $\widehat{\mathcal{H}}$ as a series in the parameter $1/n_c$, corresponding to paraxial and higher order approximations, by adopting suitably the Foldy-Wouthuysen transformation (FWT) technique applied usually to the Dirac Hamiltonian to expand it in a series of nonrelativistic and higher order relativistic approximations. It may be noted here that the FWT technique, originally developed for the Dirac electron theory, can be used to handle certain types of matrix evolution equations. The Feshbach-Villars form of the Helmholtz scalar wave equation and the matrix representation of the Maxwell equations, like in (Khan, 2005a) and the present one, fall under this class of equations which can be treated with a Foldy-Wouthuysen-like (FW-like) transformation technique which leads to successive approximations of the respective optical Hamiltonians.

Integrating the evolution equation is simple when the Hamiltonian is diagonal. So, it is desirable to find a representation in which the Hamiltonian becomes diagonal. The purpose of the FW-like transformation is to achieve this. When the medium is inhomogeneous with spatially varying refractive index it is not possible to achieve this diagonalization exactly. Thus, a series of FW-like transformations are performed leading to a series of approximate Hamiltonians which are even operators or block diagonalized. In the case of a homogeneous medium with constant refractive index it is possible to diagonalize the Hamiltonian exactly as we shall see later. A fairly detailed account of the FW-like transformation technique in optics can be found in (Khan, 2006, 2008).

Propagation of a quasiparaxial optical beam through an inhomogneous medium in which the refractive index fluctuates around a constant value can be studied using an FW-like transformation technique. We shall take the refractive index of the medium to be given by $n(\boldsymbol{r}_\perp; z) = n_c + \delta n(\boldsymbol{r}_\perp; z)$ where the constant n_c is the average value of n and $\delta n(\boldsymbol{r}_\perp; z)$ is the fluctuation. Now, let

$$\overline{\Phi}' = e^{-\frac{1}{2n_c}\mathcal{B}\widehat{\mathcal{O}}}\,\overline{\Phi}. \tag{63}$$

Then, substituting $\overline{\Phi} = e^{\frac{1}{2n_c}\mathcal{B}\widehat{\mathcal{O}}}\overline{\Phi}'$ in (54a) leads to

$$i\lambda\!\!\!^{-}\left(\frac{\partial e^{\frac{1}{2n_c}\mathcal{B}\widehat{\mathcal{O}}}}{\partial z}\right)\overline{\Phi}' + i\lambda\!\!\!^{-} e^{\frac{1}{2n_c}\mathcal{B}\widehat{\mathcal{O}}}\,\frac{\partial\overline{\Phi}'}{\partial z} = \widehat{\mathcal{H}}\left(e^{\frac{1}{2n_c}\mathcal{B}\widehat{\mathcal{O}}}\,\overline{\Phi}'\right). \tag{64}$$

Operating on both sides of this equation from left by $e^{-\frac{1}{2n_c}\mathcal{B}\widehat{\mathcal{O}}}$ and rearranging the terms we get the z-evolution equation for $\overline{\Phi}'$

$$\begin{aligned} i\lambda\!\!\!^{-}\frac{\partial\overline{\Phi}'}{\partial z} &= \widehat{\mathcal{H}}'\overline{\Phi}', \\ \widehat{\mathcal{H}}' &= \left[e^{-\frac{1}{2n_c}\mathcal{B}\widehat{\mathcal{O}}}\,\widehat{\mathcal{H}}\,e^{\frac{1}{2n_c}\mathcal{B}\widehat{\mathcal{O}}} - i\lambda\!\!\!^{-} e^{-\frac{1}{2n_c}\mathcal{B}\widehat{\mathcal{O}}}\left(\frac{\partial e^{\frac{1}{2n_c}\mathcal{B}\widehat{\mathcal{O}}}}{\partial z}\right)\right]. \end{aligned} \tag{65}$$

To calculate $\widehat{\mathcal{H}}'$ the following identities are helpful:

$$e^{-\widehat{A}}\widehat{B}e^{\widehat{A}} = \widehat{B} - [\widehat{A}, \widehat{B}] + \frac{1}{2!}\left[\widehat{A}, [\widehat{A}, \widehat{B}]\right] - \frac{1}{3!}\left[\widehat{A}, [\widehat{A}, [\widehat{A}, \widehat{B}]]\right] + \ldots, \tag{66a}$$

$$e^{-\widehat{A}}\left(\frac{\partial e^{\widehat{A}}}{\partial z}\right) = \frac{\partial \widehat{A}}{\partial z} - \frac{1}{2!}\left[\widehat{A}, \frac{\partial \widehat{A}}{\partial z}\right] + \frac{1}{3!}\left[\widehat{A}, \left[\widehat{A}, \frac{\partial \widehat{A}}{\partial z}\right]\right] - \ldots. \tag{66b}$$

Using these identities with $\widehat{A} = \mathcal{B}\widehat{\mathcal{O}}/(2n_c)$ and $\widehat{B} = \widehat{\mathcal{H}}$, we get

$$\begin{aligned} \widehat{\mathcal{H}}' &= -n_c\mathcal{B} + \widehat{\mathcal{E}}' + \widehat{\mathcal{O}}', \\ \widehat{\mathcal{E}}' &\approx \widehat{\mathcal{E}} - \mathcal{B}\left(\frac{1}{2n_c}\widehat{\mathcal{O}}^2 - \frac{1}{8n_c^3}\widehat{\mathcal{O}}^4\right) \\ &\quad - \frac{1}{8n_c^2}\left[\widehat{\mathcal{O}}, \left([\widehat{\mathcal{O}}, \widehat{\mathcal{E}}] + i\lambda\!\!\!^{-}\frac{\partial\widehat{\mathcal{O}}}{\partial z}\right)\right], \\ \widehat{\mathcal{O}}' &\approx -\frac{1}{2n_c}\mathcal{B}\left([\widehat{\mathcal{O}}, \widehat{\mathcal{E}}] + i\lambda\!\!\!^{-}\frac{\partial\widehat{\mathcal{O}}}{\partial z}\right) - \frac{1}{3n_c^2}\widehat{\mathcal{O}}^3. \end{aligned} \tag{67}$$

where $\widehat{\mathcal{E}}$ and $\widehat{\mathcal{O}}$ are as given in (54). The terms $\widehat{\mathcal{E}}'$ and $\widehat{\mathcal{O}}'$ are, respectively, even and odd, and satisfy the same relations as in (56):

$$\mathcal{B}\widehat{\mathcal{E}}' = \widehat{\mathcal{E}}'\mathcal{B}, \qquad \mathcal{B}\widehat{\mathcal{O}}' = -\widehat{\mathcal{O}}'\mathcal{B}. \tag{68}$$

Now we shall approximate $\widehat{\mathcal{H}}'$ by dropping the odd term $\widehat{\mathcal{O}}'$ and keeping only terms of first order in the order parameter $1/n_c$. The resulting Hamiltonian is

$$\widehat{\mathcal{H}}^{(2)} = \mathcal{B}\left(-n_c - \frac{1}{2n_c}\widehat{\mathcal{O}}^2\right) + \widehat{\mathcal{E}}, \tag{69}$$

where the superscript $^{(2)}$ indicates that only terms up to quadratic in $\boldsymbol{r}_\perp$ and $\nabla_\perp$ are present in it. This is the paraxial approximation and $\widehat{\mathcal{H}}^{(2)}$ is the

Maxwell vector wave optical paraxial Hamiltonian. If we call $\widehat{\mathcal{H}}^{(2)}$ as $\widehat{\mathcal{H}}_p$ and $\overline{\Phi}'$ as $\overline{\Phi}_p$, then we have, with

$$\widehat{\wp}_{\perp}^{2} = \widehat{\wp}_{x}^{2} + \widehat{\wp}_{y}^{2}, \tag{70}$$

$$i\lambda\!\!\!^{-} \frac{\partial \overline{\Phi}_p}{\partial z} = \widehat{\mathcal{H}}_p \overline{\Phi}_p, \qquad \widehat{\mathcal{H}}_p \approx \mathcal{B}\left(-n(\boldsymbol{r}_{\perp}; z) + \frac{1}{2n_c} \widehat{\wp}_{\perp}^{2} \right), \tag{71}$$

where we have assumed n and η to vary slowly such that $\nabla\bar{n}$ and $\nabla\bar{\eta}$ can be neglected in $\widehat{\mathcal{E}}$ and $\widehat{\mathcal{O}}$. Note that

$$\widehat{\mathcal{H}}_p \approx \mathcal{B}\left(-\sqrt{n^2(\boldsymbol{r}_{\perp}; z) - \widehat{\wp}_{\perp}^{2}} \right), \tag{72}$$

with

$$n(\boldsymbol{r}_{\perp}; z) = n_c + \delta n(\boldsymbol{r}_{\perp}; z), \qquad |\delta n(\boldsymbol{r}_{\perp}; z)| \ll n_c. \tag{73}$$

It should be remembered that after the final result for $\overline{\Phi}_p$ is obtained by integrating (71) for any system the fields are to be obtained by reversing the transformation and using (62):

$$\mathcal{F} = e^{-i\omega t} \mathcal{T}\left(e^{\frac{1}{2n_c}\widehat{\mathcal{O}}} \overline{\Phi}_p \right). \tag{74}$$

To go beyond the paraxial approximation we have to carry out more FW-like transformations with the same recipe as in the first transformation (63). Thus, the next step is the transformation

$$\overline{\Phi}'' = e^{-\frac{1}{2n_c}\mathcal{B}\widehat{\mathcal{O}}'} \overline{\Phi}', \tag{75}$$

which leads to

$$\begin{aligned} i\lambda\!\!\!^{-} \frac{\partial \overline{\Phi}''}{\partial z} &= \widehat{\mathcal{H}}'' \overline{\Phi}'', \\ \widehat{\mathcal{H}}'' &= \left[e^{-\frac{1}{2n_c}\mathcal{B}\widehat{\mathcal{O}}'} \widehat{\mathcal{H}}' e^{\frac{1}{2n_c}\mathcal{B}\widehat{\mathcal{O}}'} - i\lambda\!\!\!^{-} e^{-\frac{1}{2n_c}\mathcal{B}\widehat{\mathcal{O}}'} \left(\frac{\partial e^{-\frac{1}{2n_c}\mathcal{B}\widehat{\mathcal{O}}'}}{\partial z} \right) \right]. \end{aligned} \tag{76}$$

Note that the right hand sides of (65) and (76) are identical except for the replacements of $\widehat{\mathcal{E}}$ and $\widehat{\mathcal{O}}$, respectively, by $\widehat{\mathcal{E}}'$ and $\widehat{\mathcal{O}}'$. This means that we can write down $\widehat{\mathcal{H}}''$ by replacing $\widehat{\mathcal{E}}$ and $\widehat{\mathcal{O}}$, in the expression (67) for $\widehat{\mathcal{H}}'$, respectively, by $\widehat{\mathcal{E}}'$ and $\widehat{\mathcal{O}}'$. Thus, we obtain

$$\begin{aligned}
\widehat{\mathcal{H}}'' &= -n_c\mathcal{B} + \widehat{\mathcal{E}}'' + \widehat{\mathcal{O}}'', \\
\widehat{\mathcal{E}}'' &\approx \mathcal{E}' - \mathcal{B}\left(\frac{1}{2n_c}\widehat{\mathcal{O}}'^2 - \frac{1}{8n_c^3}\widehat{\mathcal{O}}'^4\right) \\
&\quad - \frac{1}{8n_c^2}\left[\widehat{\mathcal{O}}', \left([\widehat{\mathcal{O}}', \widehat{\mathcal{E}}'] + i\lambda\!\!\!^{-}\frac{\partial\widehat{\mathcal{O}}'}{\partial z}\right)\right], \\
\widehat{\mathcal{O}}'' &\approx -\frac{1}{2n_c}\mathcal{B}\left([\widehat{\mathcal{O}}', \widehat{\mathcal{E}}'] + i\lambda\!\!\!^{-}\frac{\partial\widehat{\mathcal{O}}'}{\partial z}\right) - \frac{1}{3n_c^2}\widehat{\mathcal{O}}'^3.
\end{aligned} \tag{77}$$

One can continue the process of FW-like transformations up to the desired order of accuracy. In $\widehat{\mathcal{H}}$ the odd term $\widehat{\mathcal{O}}$ is of order zero in the order parameter $1/n_c$, in $\widehat{\mathcal{H}}'$ the odd term $\widehat{\mathcal{O}}'$ is of order one in the order parameter $1/n_c$, and in $\widehat{\mathcal{H}}''$ the odd term $\widehat{\mathcal{O}}''$ is of order two in the order parameter $1/n_c$, and so on, thus showing that the odd term is weakened at each step. One can stop at any step of desired order. Stopping at the third step we shall drop the odd term $\widehat{\mathcal{O}}'''$ from $\widehat{\mathcal{H}}'''$ and write the resulting Maxwell vector wave optical Hamiltonian as

$$\begin{aligned}
\widehat{\mathcal{H}}^{(4)} &= \mathcal{B}\left(-n_c - \frac{1}{2n_c}\widehat{\mathcal{O}}^2 + \frac{1}{8n_c^3}\left\{\widehat{\mathcal{O}}^4 + \left([\widehat{\mathcal{O}}, \widehat{\mathcal{E}}] + i\lambda\!\!\!^{-}\frac{\partial\widehat{\mathcal{O}}}{\partial z}\right)^2\right\}\right) \\
&\quad + \widehat{\mathcal{E}} - \frac{1}{8n_c^2}\left[\widehat{\mathcal{O}}, \left([\widehat{\mathcal{O}}, \widehat{\mathcal{E}}] + i\lambda\!\!\!^{-}\frac{\partial\widehat{\mathcal{O}}}{\partial z}\right)\right].
\end{aligned} \tag{78}$$

This $\widehat{\mathcal{H}}^{(4)}$ is seen to contain terms up to fourth order in $\boldsymbol{r}_\perp$ and $\nabla_\perp$. Then, we will have

$$i\lambda\!\!\!^{-}\frac{\partial\overline{\Phi}'''}{\partial z} = \widehat{\mathcal{H}}^{(4)}\overline{\Phi}'''. \tag{79}$$

After integrating this equation, to get the fields corresponding to the final result, we will have to first reverse the transformations and then use (62):

$$\mathcal{F} = e^{-i\omega t}\mathcal{T}\left(e^{\frac{1}{2n_c}\widehat{\mathcal{O}}}e^{\frac{1}{2n_c}\widehat{\mathcal{O}}'}e^{\frac{1}{2n_c}\widehat{\mathcal{O}}''}\overline{\Phi}'''\right). \tag{80}$$

The matrix formalism of the Maxwell vector wave optics makes extensive use of quantum methodology (Khan, 2006, 2008, 2017a). The quantum theory of charged particle beam optics pioneered by Jagannathan et al. (Jagannathan, 1990, 1993; Jagannathan et al., 1989) had a deep impact on the matrix formulation of the Maxwell vector wave optics. Many techniques in our approach to light beam optics originate from the formalism of quantum mechanics of charged particle beam optics applicable to devices from electron microscopes to particle accelerators (Conte et al., 1996; Jagannathan, 1990, 1993, 1999, 2002, 2004; Jagannathan & Khan, 1995, 1996, 1997; Jagannathan et al., 1989; Khan, 1997, 1999a, 1999b, 2002, 2016a, 2017a, 2017b, 2018c; Khan & Jagannathan, 1995, 2020, 2021, 2024a, 2024b). A comprehensive account of the quantum theory of charged particle beam optics is available in the book titled Quantum Mechanics of Charged Particle Beam Optics: Understanding Devices from Electron Microscopes to Particle Accelerators (Jagannathan & Khan, 2019). The classic textbook of Hawkes and Kasper on Electron Optics, with eighty chapters spread over four volumes (Hawkes & Kasper, 2017a, 2017b, 2022a, 2022b), has an encyclopedic coverage of all fundamental and applied aspects of electron optics from geometrical electron optics to electron wave optics, and about half of Chapter-56, in Volume-3, is devoted to the quantum theory of electron optics based on the Dirac equation developed by us. The quantum corrections to the results of the classical theory of electron optics we have found are, of course, "...some small corrections (fortunately usually negligible) to the standard theory" as remarked by Hawkes (Hawkes, 2020).

In the context of quantum methodology in charged particle beam optics we should also note the following. Quantum methods are used to study multi-particle effects in charged particle beams. In the thermal wave model (Fedele et al., 1993; Fedele. & Man'ko, 1999; Fedele et al., 2000, 2014a, 2014b) the basic equation is a Schrödinger-like equation with the beam emittance playing the role of $\hbar$, the Planck's constant. A stochastic collective dynamical model for treating the charged particle beam phenomenologically as a quasiclassical many-body system has been developed (Petroni et al., 2000). The quantum-like approach of these models has been used to develop a diffraction model for the beam halo (Khan & Pusterla, 1999, 2000a, Khan and Pusterla, 2000b, 2001).

4. Transition to scalar wave optics and ray optics

When the time-independent electric field of a monochromatic wave $\overline{\boldsymbol{E}}(\boldsymbol{r})$ is replaced in (30) by a scalar wave function $\Psi(\boldsymbol{r})$, describing the amplitude of light at $\boldsymbol{r}$, we get the Helmholtz equation (31), the basic equation of the Helmholtz scalar wave optics. The Helmholtz equation is exact for a homogeneous medium and is approximate in the case of an inhomogeneous medium (see Appendix A). Multiplying (31) throughtout from left by $-\,\lambda\!\!\!^{-2}$ it can be rewritten as

$$\left(-\lambda\!\!\!^{-2}\nabla^2 - n^2(\boldsymbol{r})\right)\Psi(\boldsymbol{r}) = 0. \tag{81}$$

Then, a rearrangement of the terms leads to the equation

$$\left(-i\lambda\!\!\!^{-}\frac{\partial}{\partial z}\right)^2\Psi(\boldsymbol{r}) = \left(n^2(\boldsymbol{r}) - \widehat{\wp}_\perp^2\right)\Psi(\boldsymbol{r}). \tag{82}$$

At this stage, in order to write (82) as a z-evolution equation linear in $\partial/\partial z$ one resorts to the 'square-root' in the following manner (Dragt & Forest, 1986; Dragt et al., 1986):

$$i\lambda\!\!\!^{-}\frac{\partial\Psi(\boldsymbol{r})}{\partial z} = \widehat{\mathrm{H}}\,\Psi(\boldsymbol{r}), \qquad \widehat{\mathrm{H}} = \mp\sqrt{\left(n^2(\boldsymbol{r}) - \widehat{\wp}_\perp^2\right)}, \tag{83}$$

where $\widehat{\mathrm{H}}$ is the Helmholtz scalar wave optical Hamiltonian. If we consider a beam moving along the z-axis, the optic axis of a system, according to (83) the negative/positive sign corresponds to propagation along the positive/negative z-direction. It is to be noted that the beam-optical Hamiltonian in the square-root approach is identical to the one obtained using the Fermat's principle of least time (see Appendix C).

The Helmholtz scalar wave optical Hamiltonian in (83) is usually expanded in a power series in $\wp_\perp/n_c$ where n_c is the constant average value of the refractive index around which $n(\boldsymbol{r})$ fluctuates. The spatially varying refractive index $n(\boldsymbol{r})$ is also expanded to the same order consistent with the expansion of $\widehat{\mathrm{H}}$. For a homogeneous medium with a constant refractive index n_c we note the following Taylor expansion for the Hamiltonian:

$$\begin{aligned}
\widehat{\mathrm{H}}_c &= -\sqrt{\left(n_c^2 - \widehat{\wp}_\perp^2\right)} = -n_c\sqrt{\left(1 - \frac{1}{n_c^2}\widehat{\wp}_\perp^2\right)} \\
&= n_c\left(-1 + \frac{1}{2n_c^2}\widehat{\wp}_\perp^2 + \frac{1}{8n_c^4}\widehat{\wp}_\perp^4\right. \\
&\qquad \left. + \frac{1}{16n_c^6}\widehat{\wp}_\perp^6 + \frac{5}{128n_c^8}\widehat{\wp}_\perp^8 + \frac{7}{256n_c^{10}}\widehat{\wp}_\perp^{10} + \cdots\right) \\
&= n_c\sum_{m=0}^{\infty}\frac{(2m-3)\,!!}{(2m)\,!!}\frac{1}{n_c^{2m}}\widehat{\wp}_\perp^{2m},
\end{aligned} \tag{84}$$

where the !! is the double factorial with $1!! = 1$, $0!! = 1$, $(-1)!! = 1$, $(-3)!! = -1$, $(2n)!! = 2^n n!$, and $(2n-1)!! = (2n)!/(2n)!!$. The paraxial or ideal behaviour is governed by the quadratic terms in the Hamiltonian. The higher-order terms are responsible for the deviations from the ideal behaviour and are called as the aberrating terms. These terms govern the aberrations of the corresponding order. We note that even in the case of a constant refractive index, the system is inherently aberrating. It follows from (84) that for a monochromatic paraxial beam propagating predominantly in the z-direction the z-evolution equation can be taken as

$$i\lambdabar \frac{\partial \Psi(\boldsymbol{r})}{\partial z} = \widehat{\mathsf{H}}_c\, \Psi(\boldsymbol{r}), \qquad \widehat{\mathsf{H}}_c \approx -n_c + \frac{1}{2n_c}\widehat{\wp}_{\perp}^{2}. \tag{85}$$

Ray optics is the short wavelength, or high frequency, limit of the scalar wave optics. In the Hamiltonian approach to ray optics in a medium of refractive index $n(\boldsymbol{r})$ the optical phase space is taken to have the coordinates $(x(z),\, y(z),\, \wp_x = n\cos\alpha,\, \wp_y = n\cos\beta)$ where $x(z)$ and $y(z)$ are the coordinates of the point of intersection of the ray in the vertical plane at z in the optic axis (or the z-axis) and $(\cos\alpha,\ \cos\beta)$ are the direction cosines of the ray vector, with α and β being the angles the ray vector makes with the x and y axes, respectively. Note that in the representation of a plane electromagnetic wave moving in a homogeneous medium of constant refractive index n_c we have

$$\begin{aligned} &\exp\{i[n_c(k_x x + k_y y + k_z z) - \omega t]\} \\ &\quad = \exp\{i[n_c k(x\cos\alpha + y\cos\beta + z\cos\gamma) - \omega t]\}, \end{aligned} \tag{86}$$

where $\cos\alpha$, $\cos\beta$, and $\cos\gamma$ are direction cosines of the wave vector $n_c\boldsymbol{k}$, with α, β, and γ being the angles the wave vector makes with the x, y, and z axes, respectively. It is seen that

$$\begin{aligned} &\widehat{\wp}_x \exp\{i[n_c(k_x x + k_y y + k_z z) - \omega t]\} \\ &\quad = \lambdabar n_c k_x \exp\{i[n_c(k_x x + k_y y + k_z z) - \omega t]\} \\ &\quad = n_c\cos\alpha \exp\{i[n_c(k_x x + k_y y + k_z z) - \omega t]\}, \\ &\widehat{\wp}_y \exp\{i[n_c(k_x x + k_y y + k_z z) - \omega t]\} \\ &\quad = \lambdabar n_c k_y \exp\{i[n_c(k_x x + k_y y + k_z z) - \omega t]\} \\ &\quad = n_c\cos\beta \exp\{i[n_c(k_x x + k_y y + k_z z) - \omega t]\}. \end{aligned} \tag{87}$$

This suggests that we can take the ray optical Hamiltonian for a medium of constant refractive index n_c to be given by

$$H_c = -\sqrt{n_c^2 - \wp_{\perp}^2}, \tag{88}$$

as obtained from $\widehat{\mathsf{H}}_c$ in (84) replacing $\widehat{\wp}_\perp$ by $\wp_\perp$. For an inhomogeneous medium with the refractive index $n(\boldsymbol{r})$ the ray optical Hamiltonian can be taken as

$$H = -\sqrt{n(\boldsymbol{r})^2 - \wp_\perp^2}\,, \tag{89}$$

as obtained from $\widehat{\mathsf{H}}$ in (83) by replacing $(\boldsymbol{r}, \widehat{\wp}_\perp)$ by $(\boldsymbol{r}, \wp_\perp)$, respectively. This transition from the Hamiltonian scalar wave optics to the Hamiltonian ray optics is exactly like the transition from quantum mechanics to classical mechanics. It was shown by Gloge and Marcuse (1969) that the relation between ray optics and wave optics is analogous to the relation between the classical mechanics of a point particle and quantum mechanics, with $\lambda\!\!\!^{-}$ taking the place of $\hbar$ such that wave and ray optics coincide in the limit $\lambda\!\!\!^{-} \longrightarrow 0$.

Hamilton's equations of motion corresponding to the ray optical Hamiltonian H are:

$$\frac{d\boldsymbol{r}_\perp}{dz} = \frac{\partial H}{\partial \wp_\perp}, \qquad \frac{d\wp_\perp}{dz} = -\frac{\partial H}{\partial \boldsymbol{r}_\perp}. \tag{90}$$

In the present context, the Poisson bracket between any two functions $f(\boldsymbol{r}_\perp, \wp_\perp)$ and $g(\boldsymbol{r}_\perp, \wp_\perp)$ is given by

$$\{f, g\} = \left[\left(\frac{\partial f}{\partial x}\frac{\partial}{\partial \wp_x} - \frac{\partial f}{\partial \wp_x}\frac{\partial}{\partial x}\right) + \left(\frac{\partial f}{\partial y}\frac{\partial}{\partial \wp_y} - \frac{\partial f}{\partial \wp_y}\frac{\partial}{\partial y}\right)\right] g = :f:\ g, \tag{91}$$

where $:f:$ is a linear operator called the Lie operator associated with f. Note that

$$:x:\ \wp_x = :y:\ \wp_y = 1, \tag{92a}$$

$$:x:\ y = :x:\ \wp_y = :y:\ \wp_x = :\wp_x:\ \wp_y = 0, \tag{92b}$$

$$:f:\ g = -:g:\ f, \quad :fg:\ k = f\,(:g:\ k) + (:f:\ k)\,g. \tag{92c}$$

In terms of the Poisson brackets Hamilton's equations are:

$$\frac{d\boldsymbol{r}_\perp}{dz} = :-H:\ \boldsymbol{r}_\perp, \qquad \frac{d\wp_\perp}{dz} = :-H:\ \wp_\perp. \tag{93}$$

These z-evolution equations for the ray coordinates can be integrated formally leading to

$$\begin{pmatrix} \boldsymbol{r}_\perp \\ \wp_\perp \end{pmatrix}_z = \left(\mathcal{T}(z, z') \begin{pmatrix} \boldsymbol{r}_\perp \\ \wp_\perp \end{pmatrix} \right)_{z'}, \tag{94a}$$

$$\mathcal{T}(z, z') = \mathbb{P}\left[\exp\left(\int_{z'}^{z} dz : -H(z): \right) \right], \tag{94b}$$

when the plane of observation is at z and the input plane is at z'. The z-evolution operator $\mathcal{T}(z, z')$ can be expressed in an exponential form using the Magnus formula where the commutator bracket in (60) is to be replaced by the Poisson bracket.

Equation (94) is the basis of the Lie algebraic approach to ray optics. Essentially, Lie methods, used in the study of any particular optical system, are techiques to handle the computation of the evolution operator $\mathcal{T}(z, z')$ in an efficient way. Lie methods are applicable to scalar wave optics also. In the case of scalar wave optics, the evolution operator contains differential operators since the Hamiltonians are functions of $\boldsymbol{r}_\perp$ and $\widehat{\wp}_\perp$ and act on the wave function of the system. In handling such evolution operators the Poisson brackets get replaced by the commutator brackets.

Let us write the ray optical Hamiltonian for a homogneous medium of constant refractive index n_c as

$$H_c = -\sqrt{n_c^2 - \wp_\perp^2}\,. \tag{95}$$

In this case, since H_c is independent of z, the z-evolution operator is obtained exactly as

$$\begin{aligned} \mathcal{T}(z, z') &= e^{\Delta z : -H_c :} \\ &= \left(1 + \Delta z : -H_c : + \frac{(\Delta z)^2}{2!} : -H_c :^2 \right. \\ &\quad \left. + \frac{(\Delta z)^3}{3!} : -H_c :^3 + \cdots \right), \qquad \Delta z = z - z'. \end{aligned} \tag{96}$$

Using (94) and (96), we obtain the transfer map

$$\begin{pmatrix} \boldsymbol{r}_\perp(z) \\ \frac{\wp_\perp(z)}{n_c} \end{pmatrix} = \begin{pmatrix} 1 & \frac{\Delta z}{\sqrt{1 - \frac{\wp_\perp^2}{n_c^2}}} \\ 0 & 1 \end{pmatrix} \begin{pmatrix} \boldsymbol{r}_\perp(z') \\ \frac{\wp_\perp(z')}{n_c} \end{pmatrix}. \tag{97}$$

In the paraxial case the Hamiltonian can be taken as

$$H_c \approx -n_c + \frac{\wp_\perp^2}{2n_c}. \tag{98}$$

Then, Hamilton's equations (93) become

$$\frac{d\boldsymbol{r}_\perp}{dz} = \frac{\wp_\perp}{n_c}, \qquad \frac{d\wp_\perp}{dz} = 0. \tag{99}$$

Thus, for a paraxial beam

$$\frac{\wp_x}{n_c} = \frac{dx}{dz} \ll 1, \qquad \frac{\wp_y}{n_c} = \frac{dy}{dz} \ll 1, \tag{100}$$

and hence,

$$\frac{\Delta z}{\sqrt{1 - \frac{\wp_\perp^2}{n_c^2}}} \approx \Delta z. \tag{101}$$

In view of this, the transfer map (97) becomes

$$\begin{pmatrix} \boldsymbol{r}_\perp(z) \\ \frac{\wp_\perp(z)}{n_c} \end{pmatrix} = \begin{pmatrix} 1 & \Delta z \\ 0 & 1 \end{pmatrix} \begin{pmatrix} \boldsymbol{r}_\perp(z') \\ \frac{\wp_\perp(z')}{n_c} \end{pmatrix}. \tag{102}$$

Equation (97) shows that if the beam is not paraxial $\boldsymbol{r}_\perp(z)$ depends non-linearly on $\wp_\perp(z')$, in fact, depends on all powers of $\wp_\perp(z')$. Thus even the ideal system of a source-free medium of constant refractive index has aberrations to all orders (Dragt & Forest, 1986; Dragt et al., 1986). Only for a perfectly paraxial beam the transfer map (102) corresponds to the ideal straight line propagation.

5. Applications to specific systems

Now we shall look at two specific examples for the application of the matrix formalism of the Maxwell vector wave optics. The first example is that of a medium with constant refractive index and constant impedance. This system is the simplest and exactly treatable. At the same time, it enables us to have a closer look at the series expansion obtained using an FW-like iterative procedure and compare it with the exact result. Our second example is that of an axially symmetric graded-index medium. This is a well studied system using diverse techniques. We are able to reproduce the well known results which get modified by wavelength-dependent corrections. Both the paraxial behaviour and all the aberration coefficients get modified by wavelength-dependent corrections.

5.1 Medium with constant refractive index and constant impedance

The Hamiltonian governing the propagation of a beam through a homogneous medium can be actually diagonalized exactly. This is very similar to the exact diagonalization of the free particle Dirac Hamiltonian in relativistic quantum mechanics (see, e.g., Bjorken & Drell, 1994; Greiner, 2000). As in the Dirac electron theory, the other optical systems with varying refractive index can be diagonalized approximately using an FW-like iterative procedure (Khan, 2006, 2008). Hence, we have extensively used the FW-like scheme in the matrix formalism of the Maxwell vector wave optics.

Let us consider a source-free homogeneous medium of constant refractive index n_c. For such a medium $\delta n = 0$, $\nabla \bar{n} = 0$, and $\nabla \bar{\eta} = 0$, in (54). Then, the Maxwell vector wave optical z-evolution equation reads

$$i\lambda\!\!\!^{-} \frac{\partial \overline{\Phi}}{\partial z} = \widehat{\mathcal{H}}_c \overline{\Phi}, \tag{103}$$

with the Hamiltonian

$$\begin{aligned} \widehat{\mathcal{H}}_c &= -n_c \mathcal{B} + \widehat{\mathcal{O}}, \\ \widehat{\mathcal{O}} &= \begin{pmatrix} 0_2 & 0_2 & 1\!\!1_2 \widehat{\wp}_- & 0_2 \\ 0_2 & 0_2 & 0_2 & 1\!\!1_2 \widehat{\wp}_+ \\ -1\!\!1_2 \widehat{\wp}_+ & 0_2 & 0_2 & 0_2 \\ 0_2 & -1\!\!1_2 \widehat{\wp}_- & 0_2 & 0_2 \end{pmatrix}. \end{aligned} \tag{104}$$

Let us now make an FW-like transformation

$$\begin{aligned} \overline{\Phi}_c &= \widehat{\mathsf{T}}_c \overline{\Phi} = \begin{pmatrix} \overline{\phi}_c^{\,+} \\ \overline{\phi}_c^{\,-} \end{pmatrix}, \\ \widehat{\mathsf{T}}_c &= e^{-\theta \mathcal{B} \widehat{\mathcal{O}}}, \qquad \theta = \frac{1}{2 \mid \widehat{\wp}_\perp \mid} \tanh^{-1} \frac{\mid \widehat{\wp}_\perp \mid}{n_c}. \end{aligned} \tag{105}$$

The z-evolution equation for $\overline{\Phi}_c$ becomes

$$i\lambda\!\!\!^{-} \frac{\partial \overline{\Phi}_c}{\partial z} = \left(\widehat{\mathsf{T}}_c \widehat{\mathcal{H}}_c \widehat{\mathsf{T}}_c^{-1} \right) \overline{\Phi}_c, \tag{106}$$

since $\widehat{\mathsf{T}}_c$ is independent of z. Now, it is found that we can get the exact forms of $\widehat{\mathsf{T}}_c$ and $\widehat{\mathsf{T}}_c^{-1}$:

$$\begin{aligned}\widehat{\mathsf{T}}_c &= 1\!\!1_2 \cosh\left(\theta\,|\widehat{\wp}_\perp|\right) - \mathcal{B}\widehat{\mathcal{O}}\frac{1}{|\widehat{\wp}_\perp|}\sinh\left(\theta\,|\widehat{\wp}_\perp|\right),\\ &= \frac{\left(1\!\!1_8\widehat{\wp}_z - \mathcal{B}\widehat{\mathcal{H}}_c\right)}{\sqrt{2\widehat{\wp}_z\left(n_c+\widehat{\wp}_z\right)}},\end{aligned} \tag{107}$$

and

$$\widehat{\mathsf{T}}_c^{-1} = \frac{\left(1\!\!1_8\widehat{\wp}_z - \widehat{\mathcal{H}}_c\mathcal{B}\right)}{\sqrt{2\widehat{\wp}_z\left(n_c+\widehat{\wp}_z\right)}}, \tag{108}$$

where

$$\widehat{\wp}_z = \sqrt{n_c^2 - \left(\widehat{\wp}_x^2 + \widehat{\wp}_y^2\right)} = \sqrt{n_c^2 - \widehat{\wp}_\perp^2}. \tag{109}$$

Then, it follows that

$$\widehat{\mathsf{T}}_c\widehat{\mathcal{H}}_c\widehat{\mathsf{T}}_c^{-1} = \widehat{\mathcal{H}}_c^{\mathrm{diag}} = -\mathcal{B}\widehat{\wp}_z, \tag{110}$$

and hence the z-evolution equation for $\overline{\Phi}_c$, (106), becomes

$$i\lambda\!\!\!^{-}\frac{\partial\overline{\Phi}_c}{\partial z} = \widehat{\mathcal{H}}_c^{\mathrm{diag}}\overline{\Phi}_c = -\mathcal{B}\widehat{\wp}_z\overline{\Phi}_c. \tag{111}$$

For practical purposes, we can consider $\widehat{\wp}_z$ to be given by the approximation

$$\widehat{\wp}_z \approx n_c\left(1 - \frac{1}{2n_c^2}\widehat{\wp}_\perp^2 - \frac{1}{8n_c^4}\widehat{\wp}_\perp^4\right). \tag{112}$$

Since the right hand side of (111) is independent of z, it follows that

$$\begin{aligned}-\lambda\!\!\!^{-2}\frac{\partial^2\overline{\Phi}_c}{\partial z^2} &= \widehat{\wp}_z^2\overline{\Phi}_c = \left(n_c^2 - \widehat{\wp}_\perp^2\right)\overline{\Phi}_c\\ &= \left(n_c^2 + \lambda\!\!\!^{-2}\nabla_\perp^2\right)\overline{\Phi}_c.\end{aligned} \tag{113}$$

Dividing both sides of this equation by $\lambda\!\!\!^{-2}$ and rearranging the terms we get

$$\left(\nabla^2 + n_c^2k^2\right)\overline{\Phi}_c = 0, \tag{114}$$

the Helmholtz equation, for a source-free homogeneous medium.

Equation (111) shows that any of the four components of $\overline{\phi}_c^{\,+}$, say, $\overline{\phi}_{cj}^{\,+}$, obeys, independently, the scalar equation

$$i\lambda\!\!\!^{-}\frac{\partial \overline{\phi}_{cj}^{\,+}}{\partial z} = -\widehat{\wp}_z \overline{\phi}_{cj}^{\,+}, \tag{115}$$

besides the Helmholtz equation. Similarly, any of the four components of $\overline{\phi}_c^{\,-}$, say, $\overline{\phi}_{cj}^{\,-}$, obeys, independently, the scalar equation

$$i\lambda\!\!\!^{-}\frac{\partial \overline{\phi}_{cj}^{\,-}}{\partial z} = \widehat{\wp}_z \overline{\phi}_{cj}^{\,-}, \tag{116}$$

besides the Helmholtz equation. For a paraxial beam moving predominantly along the positive z direction we can take

$$\overline{\phi}_{cj}^{\,+}(\boldsymbol{r}_\perp; z) = \overline{\varphi}_{cj}^{\,+}(\boldsymbol{r}_\perp; z)\, e^{in_c kz}, \tag{117}$$

where $k = 1/\lambda\!\!\!^{-}$ is the wave vector in vacuum and $n_c k$ is the wave vector in the medium. We can take in (115)

$$-\widehat{\wp}_z \approx -n_c + \frac{1}{2n_c}\widehat{\wp}_\perp^{\,2}. \tag{118}$$

Then, the z-evolution equation for $\overline{\varphi}_{cj}^{\,+}$ becomes

$$\nabla_\perp^2\, \overline{\varphi}_{cj}^{\,+} + 2in_c k\frac{\partial \overline{\varphi}_{cj}^{\,+}}{\partial z} = 0, \tag{119}$$

the paraxial Helmholtz equation, or the paraxial approximation of the Helmholtz equation. Similarly, for a paraxial beam moving predominantly along the negative z direction we can take

$$\overline{\phi}_{cj}^{\,-}(\boldsymbol{r}_\perp; z) = \overline{\varphi}_{cj}^{\,-}(\boldsymbol{r}_\perp; z)\, e^{-in_c kz}. \tag{120}$$

Then, the z-evolution equation for $\overline{\varphi}_{cj}^{\,-}$ becomes the paraxial equation

$$\nabla_\perp^2\, \overline{\varphi}_{cj}^{\,-} - 2in_c k\frac{\partial \overline{\varphi}_{cj}^{\,-}}{\partial z} = 0, \tag{121}$$

Writing the paraxial Helmholtz equation (119) as

$$i\frac{\partial \overline{\varphi}_{cj}^{\,+}}{\partial z} = -\frac{1}{2n_c k}\nabla_\perp^2\, \overline{\varphi}_{cj}^{\,+}, \tag{122}$$

it follows from (57) and (60) that

$$\overline{\phi}_{cj}^{\,+}(\boldsymbol{r}_\perp; z) = e^{in_c k\Delta z}\exp\left(\frac{i\Delta z}{2n_c k}\nabla_\perp^2\right)\overline{\phi}_{cj}^{\,+}(\boldsymbol{r}; z'), \qquad \Delta z = z - z'. \tag{123}$$

Following (61) we can write

$$\overline{\phi}_{cj}^{+}(\boldsymbol{r}_\perp;\, z) = \int d^2r'_\perp \quad K_c^{+}(\boldsymbol{r}_\perp,\, z;\, \boldsymbol{r}'_\perp,\, z')\,\overline{\phi}_{cj}^{+}(\boldsymbol{r}'_\perp;\, z'). \tag{124}$$

To get the expression for $K_c^{+}(\boldsymbol{r}_\perp,\, z;\, \boldsymbol{r}'_\perp,\, z')$ let us proceed as follows: with $\Delta z = z - z'$,

$$\begin{aligned}
\overline{\phi}_{cj}^{+}(\boldsymbol{r}_\perp;\, z) &= e^{in_c k\Delta z}\exp\left(\frac{i\Delta z}{2n_c k}\nabla_\perp^2\right)\overline{\phi}_{cj}^{+}(\boldsymbol{r}_\perp;\, z') \\
&= \frac{e^{in_c k\Delta z}}{2\pi}\exp\left(\frac{i\Delta z}{2n_c k}\nabla_\perp^2\right)\int d^2k'_\perp \exp\left(i\boldsymbol{k}'_\perp\cdot\boldsymbol{r}_\perp\right)F\left(\boldsymbol{k}'_\perp;\, z'\right) \\
&= \frac{e^{in_c k\Delta z}}{2\pi}\int d^2k'_\perp \exp\left(-\frac{i\Delta z}{2n_c k}k_\perp'^{\,2}\right)\exp\left(i\boldsymbol{k}'_\perp\cdot\boldsymbol{r}_\perp\right)F\left(\boldsymbol{k}'_\perp;\, z'\right) \\
&= \frac{e^{in_c k\Delta z}}{4\pi^2}\int d^2k'_\perp \exp\left(-\frac{i\Delta z}{2n_c k}k_\perp'^{\,2}\right)\exp\left(i\boldsymbol{k}'_\perp\cdot\boldsymbol{r}_\perp\right) \\
&\quad\times \int d^2r'_\perp \exp\left(-i\boldsymbol{k}'_\perp\cdot\boldsymbol{r}'_\perp\right)\overline{\phi}_{cj}^{+}(\boldsymbol{r}'_\perp;\, z') \qquad (125) \\
&= \frac{e^{in_c k\Delta z}}{4\pi^2}\int d^2r'_\perp \int d^2k'_\perp \exp\left(-\frac{i\Delta z}{2n_c k}k_\perp'^{\,2}\right) \\
&\quad\times \exp\left(i\boldsymbol{k}'_\perp\cdot(\boldsymbol{r}_\perp - \boldsymbol{r}'_\perp)\right)\overline{\phi}_{cj}^{+}(\boldsymbol{r}'_\perp;\, z') \\
&= e^{in_c k\Delta z}\int d^2r'_\perp \left[\frac{n_c k}{i2\pi\Delta z}\exp\left(\frac{in_c k\,|\,\boldsymbol{r}_\perp - \boldsymbol{r}'_\perp\,|^2}{2\Delta z}\right)\right]\overline{\phi}_{cj}^{+}(\boldsymbol{r}'_\perp;\, z') \\
&= \frac{n_c e^{in_c k\Delta z}}{i\lambda\Delta z}\int d^2r'_\perp \exp\left(\frac{i\pi n_c\,|\,\boldsymbol{r}_\perp - \boldsymbol{r}'_\perp\,|^2}{\lambda\Delta z}\right)\overline{\phi}_{cj}^{+}(\boldsymbol{r}'_\perp;\, z').
\end{aligned}$$

Thus, propagation of a paraxial beam along the positive z-direction in a source-free homogeneous medium is given by

$$\begin{aligned}
\overline{\phi}_{cj}^{+}(\boldsymbol{r}_\perp;\, z) &= \int d^2r'_\perp \quad K_c^{+}(\boldsymbol{r}_\perp,\, z;\, \boldsymbol{r}'_\perp,\, z')\,\overline{\phi}_{cj}^{+}(\boldsymbol{r}'_\perp;\, z'), \\
K_c^{+}(\boldsymbol{r}_\perp,\, z;\, \boldsymbol{r}'_\perp,\, z') &= \frac{n_c e^{in_c k(z-z')}}{i\lambda(z-z')}\exp\left(\frac{i\pi n_c\,|\,\boldsymbol{r}_\perp - \boldsymbol{r}'_\perp\,|^2}{\lambda(z-z')}\right).
\end{aligned} \tag{126}$$

This result, the Fresnel integral or the Fresnel transform, is the basis of Fourier optics (see, e.g., Goodman, 1996). In the above derivation we have used the integral

$$\int_{-\infty}^{\infty} d\psi \exp\left(-a\xi^2 + i\xi\eta\right) = \sqrt{\frac{\pi}{a}}\exp\left(-\frac{\eta^2}{4a}\right). \tag{127}$$

For $\overline{\varphi}_c^{-}$ we have

$$i\frac{\partial\overline{\varphi}_{cj}^{-}}{\partial z} = \frac{1}{2n_c k}\nabla_\perp^2\,\overline{\varphi}_{cj}^{-}, \tag{128}$$

and hence

$$\overline{\phi}_{cj}^{-}(\boldsymbol{r}_{\perp};\, z) = e^{-in_c k\Delta z}\exp\left(-\frac{i\Delta z}{2n_c k}\nabla_{\perp}^{2}\right)\overline{\phi}_{cj}^{-}(\boldsymbol{r};\, z'), \quad \Delta z = (z - z'). \tag{129}$$

Consequently, we get

$$\begin{aligned} \overline{\phi}_{cj}^{-}(\boldsymbol{r}_{\perp};\, z) &= \int d^2r'_{\perp}\, K_c^{-}(\boldsymbol{r}_{\perp},\, z;\, \boldsymbol{r}'_{\perp},\, z')\,\overline{\phi}_{cj}^{-}(\boldsymbol{r}'_{\perp};\, z') \\ K_c^{-}(\boldsymbol{r}_{\perp},\, z;\, \boldsymbol{r}'_{\perp},\, z') &= \frac{in_c e^{-in_c k(z-z')}}{\lambda(z-z')}\exp\left(-\frac{i\pi n_c \mid \boldsymbol{r}_{\perp} - \boldsymbol{r}'_{\perp} \mid^2}{\lambda(z-z')}\right). \end{aligned} \tag{130}$$

From (126) and (130) we can write

$$\overline{\Phi}_c(\boldsymbol{r}_{\perp};\, z) = \int d^2r'_{\perp} \quad K_c(\boldsymbol{r}_{\perp},\, z;\, \boldsymbol{r}'_{\perp},\, z')\,\overline{\Phi}_c(\boldsymbol{r}'_{\perp};\, z'), \tag{131}$$

in which the PSF, $K_c(\boldsymbol{r}_{\perp}, z; \boldsymbol{r}'_{\perp}, z')$, is an eight dimensional diagonal matrix with $K_c^{+}(\boldsymbol{r}_{\perp}, z; \boldsymbol{r}'_{\perp}, z')$ as the four upper diagonal elements and $K_c^{-}(\boldsymbol{r}_{\perp}, z; \boldsymbol{r}'_{\perp}, z')$ as the four lower diagonal elements. From $\overline{\Phi}_c(z)$ one can obtain the fields using first the inverse transformation to $\overline{\Phi}(z)$ and then (62):

$$\mathcal{F} = e^{-i\omega t}\mathcal{T}\left(\hat{\mathsf{T}}_c^{-1}\overline{\Phi}_c\right). \tag{132}$$

Let us now compare (122) with the nonrelativistic Schrödinger equation for a free particle of mass m moving in the xy-plane:

$$i\hbar\frac{\partial\Psi(\boldsymbol{r}_{\perp};\, t)}{\partial t} = -\frac{\hbar^2}{2m}\nabla_{\perp}^{2}\,\Psi(\boldsymbol{r}_{\perp};\, t), \tag{133}$$

or,

$$i\frac{\partial\Psi(\boldsymbol{r}_{\perp};\, t)}{\partial t} = -\frac{\hbar}{2m}\nabla_{\perp}^{2}\,\Psi(\boldsymbol{r}_{\perp};\, t). \tag{134}$$

The correspondence

$$z \leftrightarrow t, \qquad n_c k \leftrightarrow m/\hbar, \tag{135}$$

is clear. From (117) and (126) it follows that

$$\begin{aligned} \overline{\varphi}_{cj}^{+}(\boldsymbol{r}_{\perp};\, z) &= \int d^2r'_{\perp} \quad \kappa_c^{+}(\boldsymbol{r}_{\perp},\, z;\, \boldsymbol{r}'_{\perp},\, z')\,\overline{\varphi}_{cj}^{+}(\boldsymbol{r}'_{\perp};\, z'), \\ \kappa_c^{+}(\boldsymbol{r}_{\perp},\, z;\, \boldsymbol{r}'_{\perp},\, z') &= \frac{n_c k}{i2\pi(z-z')}\exp\left(\frac{in_c k \mid \boldsymbol{r}_{\perp} - \boldsymbol{r}'_{\perp} \mid^2}{2(z-z')}\right), \end{aligned} \tag{136}$$

where we have replaced $1/\lambda$ by $k/2\pi$ in (126). Now, using the correspondence (135) we get

$$\begin{aligned} \Psi(\boldsymbol{r}_\perp; t) &= \int d^2 r_\perp' \quad \mathrm{K}(\boldsymbol{r}_\perp, t; \boldsymbol{r}_\perp', t')\Psi(\boldsymbol{r}_\perp'; t'), \\ \mathrm{K}(\boldsymbol{r}_\perp, t; \boldsymbol{r}_\perp', t') &= \frac{m}{2\pi i\hbar(t-t')} \exp\left(-\frac{m \mid \boldsymbol{r}_\perp - \boldsymbol{r}_\perp' \mid^2}{2i\hbar(t-t')}\right), \end{aligned} \tag{137}$$

where $\mathrm{K}(\boldsymbol{r}_\perp, t; \boldsymbol{r}_\perp', t')$ is the well known propagator for a free particle moving in two dimensions (see, e.g., Greiner, 2001). This reminds us of Hamilton's optical-mechanical analogy.

For a source-free homogeneous medium, with a constant refractive index n_c, $\widehat{\mathcal{E}} = 0$, $\partial\widehat{\mathcal{O}}/\partial z = 0$, and $\widehat{\mathcal{O}}^2 = -\widehat{\wp}_\perp^2$, the Maxwell vector wave optical Hamiltonian for the propagation of a paraxial beam becomes, as seen from (69),

$$\widehat{\mathcal{H}}_c^{(2)} = \mathcal{B}\left[-n_c\left(1 - \frac{1}{2n_c^2}\widehat{\wp}_\perp^2\right)\right]. \tag{138}$$

For the propagation of a quasiparaxial beam the Hamiltonian is

$$\widehat{\mathcal{H}}_c^{(4)} = \mathcal{B}\left[-n_c\left(1 - \frac{1}{2n_c^2}\widehat{\wp}_\perp^2 - \frac{1}{8n_c^4}\widehat{\wp}_\perp^4\right)\right], \tag{139}$$

as seen from (78). This series expansion obtained using the FW-like transformation procedure is to be compared with the exact result in (105):

$$\begin{aligned} \widehat{\mathcal{H}}_c^{\mathrm{diag}} &= \mathcal{B}\left(-\sqrt{n_c^2 - \widehat{\wp}_\perp^2}\right) = \mathcal{B}\left(-n_c\sqrt{1 - \frac{1}{n_c^2}\widehat{\wp}_\perp^2}\right) \\ &\approx \mathcal{B}\left[-n_c\left(1 - \frac{1}{2n_c^2}\widehat{\wp}_\perp^2 - \frac{1}{8n_c^4}\widehat{\wp}_\perp^4\right)\right] = \widehat{\mathcal{H}}_c^{(4)}. \end{aligned} \tag{140}$$

5.2 Axially symmetric graded-index medium

In the first example, we considered the ideal system of a homogeneous medium of constant refractive index. This system is exactly treatable. We also did it approximately in order to illustrate the FW series expansion scheme. The second example, we shall consider now, is that of an axially symmetric graded-index medium, which can not be analysed exactly. We shall take the refractive index of this system to be given by an infinite series (see, e.g., Dragt et al., 1986),

$$n(\boldsymbol{r}_\perp; z) = n_0 + \alpha_2(z) r_\perp^2 + \alpha_4(z) r_\perp^4 + \cdots . \tag{141}$$

The optic axis of the system is along the z-axis. Then, we have in (54)

$$\begin{aligned}
\widehat{\mathcal{E}} &= -(\alpha_2(z) r_\perp^2 + \alpha_4(z) r_\perp^4 + \cdots)\mathcal{B} \\
&\quad + i\lambdabar \begin{pmatrix} \boldsymbol{\sigma}\cdot\nabla\overline{n} & 0_2 & 0_2 & 0_2 \\ 0_2 & \boldsymbol{\sigma}^*\cdot\nabla\overline{n} & 0_2 & 0_2 \\ 0_2 & 0_2 & -\boldsymbol{\sigma}\cdot\nabla\overline{n} & 0_2 \\ 0_2 & 0_2 & 0_2 & -\boldsymbol{\sigma}^*\cdot\nabla\overline{n} \end{pmatrix}, \\
\widehat{\mathcal{O}} &= \begin{pmatrix} 0_2 & 0_2 & 1\!\!1_2\widehat{\wp}_- & 0_2 \\ 0_2 & 0_2 & 0_2 & 1\!\!1_2\widehat{\wp}_+ \\ -1\!\!1_2\widehat{\wp}_+ & 0_2 & 0_2 & 0_2 \\ 0_2 & -1\!\!1_2\widehat{\wp}_- & 0_2 & 0_2 \end{pmatrix}.
\end{aligned} \tag{142}$$

After three FW-like transformations involving considerable, but straightforward, algebra we arrive at the z-evolution equation

$$i\lambdabar \frac{\partial\overline{\Phi}(\boldsymbol{r}_\perp; z)}{\partial z} = \widehat{\mathcal{H}}(\boldsymbol{r}_\perp, \widehat{\wp}_\perp, z)\overline{\Phi}(\boldsymbol{r}_\perp; z), \tag{143}$$

where the Maxwell vector wave optical Hamiltonian is seen to be, up to fourth order in $(\boldsymbol{r}_\perp, \widehat{\wp}_\perp)$,

$$\begin{aligned}
\widehat{\mathcal{H}} &= \widehat{\mathcal{H}}_{0,p} + \widehat{\mathcal{H}}_{0,(4)} + \widehat{\mathcal{H}}^{(\lambdabar)}_{0,(2)} + \widehat{\mathcal{H}}^{(\lambdabar)}_{0,(4)} + \widehat{\mathcal{H}}^{(\lambdabar,\sigma)}_{(0)} + \widehat{\mathcal{H}}^{(\lambdabar,\sigma)}_{(2)} + \cdots, \\
\widehat{\mathcal{H}}_{0,p} &= \left(-n_0 + \tfrac{1}{2n_0}\widehat{\wp}_\perp^2 - \alpha_2(z) r_\perp^2\right)\mathcal{B}, \\
\widehat{\mathcal{H}}_{0,(4)} &= \left(\tfrac{1}{8n_0^3}\widehat{\wp}_\perp^4 - \tfrac{\alpha_2(z)}{4n_0^2}\left(r_\perp^2\widehat{\wp}_\perp^2 + \widehat{\wp}_\perp^2 r_\perp^2\right) - \alpha_4(z) r_\perp^4\right)\mathcal{B}, \\
\widehat{\mathcal{H}}^{(\lambdabar)}_{0,(2)} &= \left(-\tfrac{\lambdabar^2}{2n_0^2}\alpha_2(z) + \tfrac{\lambdabar^2}{2n_0^3}\alpha_2^2(z) r_\perp^2\right)\mathcal{B} - \tfrac{\lambdabar}{2n_0^2}\alpha_2(z)\widehat{L}_z\begin{pmatrix}\beta & 0_4 \\ 0_4 & \beta\end{pmatrix}, \\
\widehat{\mathcal{H}}^{(\lambdabar)}_{0,(4)} &= \left(\tfrac{\lambdabar^2}{2n_0^3}\alpha_2(z)\alpha_4(z) r_\perp^4\right)\mathcal{B} + \tfrac{\lambdabar}{4n_0^3}\alpha_2^2(z)\left(r_\perp^2\widehat{L}_z + \widehat{L}_z r_\perp^2\right)\begin{pmatrix}\beta & 0_4 \\ 0_4 & \beta\end{pmatrix}, \\
\widehat{\mathcal{H}}^{(\lambdabar,\sigma)}_{(0)} &= i\lambdabar \begin{pmatrix} \boldsymbol{\sigma}\cdot\nabla\overline{n} & 0_2 & 0_2 & 0_2 \\ 0_2 & \boldsymbol{\sigma}^*\cdot\nabla\overline{n} & 0_2 & 0_2 \\ 0_2 & 0_2 & -\boldsymbol{\sigma}\cdot\nabla\overline{n} & 0_2 \\ 0_2 & 0_2 & 0_2 & -\boldsymbol{\sigma}^*\cdot\nabla\overline{n} \end{pmatrix}, \\
\widehat{\mathcal{H}}^{(\lambdabar,\sigma)}_{(2)} &= i\tfrac{\lambdabar}{8n_0^2}\begin{pmatrix} \widehat{\Upsilon} & 0_2 & 0_2 & 0_2 \\ 0_2 & \widehat{\Upsilon}^* & 0_2 & 0_2 \\ 0_2 & 0_2 & -\widehat{\Upsilon} & 0_2 \\ 0_2 & 0_2 & 0_2 & -\widehat{\Upsilon}^* \end{pmatrix},
\end{aligned} \tag{144}$$

with β as the Dirac beta matrix, $\widehat{\Upsilon} = [\widehat{\wp}_{\perp}^{2}, \boldsymbol{\sigma}\cdot\nabla\bar{n}]_{+}$, $[A, B]_{+} = (AB + BA)$, and

$$\widehat{L}_{z} = x\widehat{\wp}_{y} - y\widehat{\wp}_{x}, \tag{145}$$

and '…' denoting the numerous other terms arising from the matrix terms. We have retained only the leading order of such terms above for an illustration.

5.2.1 Paraxial Hamiltonian

It follows from (144) that the Maxwell vector wave optical paraxial Hamiltonian for the axially symmetric graded-index medium considered is

$$\widehat{\mathcal{H}}_{p} = \widehat{\mathcal{H}}_{0,p} + \widehat{\mathcal{H}}_{0,(2)}^{(\lambda\!\!\!^{-})} + \widehat{\mathcal{H}}_{(0)}^{(\lambda\!\!\!^{-},\sigma)} + \widehat{\mathcal{H}}_{(2)}^{(\lambda\!\!\!^{-},\sigma)}. \tag{146}$$

The scalar wave optical paraxial Hamiltonian, obtained from $\widehat{\mathcal{H}}_{p}$ by dropping the matrices, is given by

$$\begin{aligned}
\widehat{\mathsf{H}}_{p} &= -n_{0} + \frac{1}{2n_{0}}\widehat{\wp}_{\perp}^{2} - \alpha_{2}(z)r_{\perp}^{2} \\
&\quad - \frac{\lambda\!\!\!^{-2}}{2n_{0}^{2}}\alpha_{2}(z) + \frac{\lambda\!\!\!^{-2}}{2n_{0}^{3}}\alpha_{2}^{2}(z)r_{\perp}^{2} - \frac{\lambda\!\!\!^{-}}{2n_{0}^{2}}\alpha_{2}(z)\widehat{L}_{z} \\
&= -n_{0}\left(1 + \frac{\lambda\!\!\!^{-2}}{2n_{0}^{3}}\alpha_{2}(z)\right) + \frac{1}{2n_{0}}\widehat{\wp}_{\perp}^{2} - \left(\alpha_{2}(z) - \frac{\lambda\!\!\!^{-2}}{2n_{0}^{3}}\alpha_{2}^{2}(z)\right)r_{\perp}^{2} \\
&\quad - \frac{\lambda\!\!\!^{-}}{2n_{0}^{2}}\alpha_{2}(z)\widehat{L}_{z},
\end{aligned} \tag{147}$$

in which only terms up to quadratic in $(\boldsymbol{r}_{\perp}, \widehat{\wp}_{\perp})$ have been retained.

It should be noted that the scalar wave optics obtained from the Maxwell vector wave optics by dropping the matrices (Note: not the 'matrix terms') is not the same as the traditional Helmholtz scalar wave optics. In the traditional Helmholtz scalar wave optics the Hamiltonian is given by $-\sqrt{(n^{2}(\boldsymbol{r}) - \widehat{\wp}_{\perp}^{2})}$ which is usually expanded as a power series in $\boldsymbol{r}_{\perp}$ and $\widehat{\wp}_{\perp}$. We shall call the scalar wave optics obtained from the Maxwell vector wave optics by dropping the matrices as the scalar wave optics subduced by the Maxwell vector wave optics. This is in analogy with the subduced representation in group theory. Let $G = \{g\}$ be a group and $H = \{h\}$ be a subgroup of G. If $\Gamma(G) = \{\Gamma(g); g \in G\}$ is a representation of G then it provides a representation of H given by $\Gamma(H) = \{\Gamma(h); h \in H\}$ called the representation of H subduced by the representation $\Gamma(G)$ of G.

Similarly, if we drop the matrices in a Maxwell vector wave optical Hamiltonian we obtain a scalar wave optical Hamiltonian subduced by the Maxwell vector wave optical Hamiltonian. The scalar wave optical Hamiltonian subduced by the Maxwell vector wave optical Hamiltonian of a system will not be the same as the traditional Helmholtz scalar wave optical Hamiltonian of the system. The difference manifests in the subduced scalar wave optical Hamiltonian as the presence of wavelength-dependent corrections to the Helmholtz scalar wave optical Hamiltonian. In the limit $\lambdabar \longrightarrow 0$ the scalar wave optics subduced by the Maxwell vector wave optics becomes the Helmholtz scalar wave optics.

From the scalar wave optical paraxial Hamiltonian $\widehat{\mathsf{H}}_p$ we obtain the ray optical paraxial Hamiltonian, using the replacement rule $(\boldsymbol{r}_\perp, \widehat{\wp}_\perp) \longrightarrow (\boldsymbol{r}_\perp, \wp_\perp)$, and taking the short wavelength limit $\lambdabar \longrightarrow 0$. Thus, we get the ray optical paraxial Hamiltonian:

$$H_p = -\left(n_0 + \alpha_2(z)\, r_\perp^2\right) + \frac{1}{2n_0}\wp_\perp^2 \tag{148}$$

Note that

$$H_p \approx -\sqrt{\left(n_0 + \alpha_2(z)\, r_\perp^2\right)^2 - \wp_\perp^2}\,, \tag{149}$$

the traditional paraxial Hamiltonian in ray optics (Dragt et al., 1986) in which the refractive index of the medium has been approximated as

$$n(\boldsymbol{r}_\perp; z) \approx n_0 + \alpha_2(z)\, r_\perp^2, \tag{150}$$

consistent with the paraxial approximation.

The scalar wave function $\Psi_p(\boldsymbol{r}_\perp; z)$ representing the paraxial beam moving through the graded-index medium satisfies the z-evolution equation

$$i\lambdabar \frac{\partial \Psi_p(z)}{\partial z} = \widehat{\mathsf{H}}_p \Psi_p(z). \tag{151}$$

Integrating this equation we can write

$$\begin{aligned} \Psi_p(z) &= \widehat{\mathcal{T}}_p(z, z')\,\Psi_p(z'), \\ \widehat{\mathcal{T}}_p(z, z') &= \mathbb{P}\left[\exp\left(-\tfrac{i}{\lambdabar}\int_{z'}^{z} dz\; \widehat{\mathsf{H}}_p(z)\right)\right], \end{aligned} \tag{152}$$

where $\mathbb{P}[\cdots]$ is a path-ordered integral a in (57) and $\widehat{\mathcal{T}}_p(z, z')$ can be expressed as an ordinary exponential using the Magnus formula (60). In integral form, we can write (152) as

$$\Psi_p(\boldsymbol{r}_\perp; z) = \int d^2 r'_\perp \quad K(\boldsymbol{r}_\perp, z; \boldsymbol{r}'_\perp, z')\Psi_p(\boldsymbol{r}'_\perp; z'). \tag{153}$$

where $K(\boldsymbol{r}_\perp, z; \boldsymbol{r}'_\perp, z')$ is the PSF of the system which contains all the information about the optical behaviour of the system including diffraction.

In the case of free propagation of a paraxial Maxwell beam in a homogeneous medium we could calculate $K_c(\boldsymbol{r}_\perp, z; \boldsymbol{r}'_\perp, z')$, the PSF, exactly (see (131)), which explains the Fresnel and Fraunhofer diffraction patterns and forms the basis of Fourier optics (see, e.g., Goodman, 1996). We also used the ray optics picture to understand the straight line path of a paraxial beam moving in the homogeneous medium. In the present case of axially symmetric graded-index medium it is not so easy to find and analyse the PSF. Hence, to understand the behaviour of a paraxial beam propagating in an axially symmetric graded-index medium we shall adapt a hybrid approach. We shall make the replacement $(\boldsymbol{r}_\perp, \widehat{\wp}_\perp) \longrightarrow (\boldsymbol{r}_\perp, \wp_\perp)$ in the scalar wave optical Hamiltonian $\widehat{\mathsf{H}}_p$ getting

$$\begin{aligned} \widetilde{\mathsf{H}}_p &= -n_0\left(1 + \frac{\lambda\!\!\!^{-2}}{2n_0^3}\alpha_2(z)\right) + \frac{1}{2n_0}\wp_\perp^2 - \left(\alpha_2(z) - \frac{\lambda\!\!\!^{-2}}{2n_0^3}\alpha_2^2(z)\right) r_\perp^2 \\ &\quad - \frac{\lambda\!\!\!^{-}}{2n_0^2}\alpha_2(z)\, L_z. \end{aligned} \tag{154}$$

We shall regard $\widetilde{\mathsf{H}}_p$, *mathematically*, as the ray optical Hamiltonian of the paraxial beam moving in the axially symmetric graded-index medium, though a ray optical Hamiltonian should not contain $\lambda\!\!\!^{-}$ since by definition ray optics is the zero wavelength limit of wave optics. With this understanding, the phase space transfer map for the paraxial system with the Hamiltonian $\widetilde{\mathsf{H}}_p$ is seen to be given by

$$\begin{pmatrix} x \\ y \\ \frac{\wp_x}{n_0} \\ \frac{\wp_y}{n_0} \end{pmatrix}_z = \left(\mathbb{P}\left[\exp\left(\int_{z'}^{z} dz \, : -\widetilde{\mathsf{H}}_p(z):\right)\right] \begin{pmatrix} x \\ y \\ \frac{\wp_x}{n_0} \\ \frac{\wp_y}{n_0} \end{pmatrix}\right)_{z'}. \tag{155}$$

Hamilton's equations corresponding to $\widetilde{\mathsf{H}}_p$ are

$$\frac{d}{dz}\begin{pmatrix} x \\ y \\ \frac{\wp_x}{n_0} \\ \frac{\wp_y}{n_0} \end{pmatrix} = :-\widetilde{\mathsf{H}}_p: \begin{pmatrix} x \\ y \\ \frac{\wp_x}{n_0} \\ \frac{\wp_y}{n_0} \end{pmatrix} = \widetilde{\mathsf{T}}\begin{pmatrix} x \\ y \\ \frac{\wp_x}{n_0} \\ \frac{\wp_y}{n_0} \end{pmatrix}, \tag{156a}$$

$$\widetilde{\mathsf{T}} = \begin{pmatrix} 0 & -\mathrm{b}(z) & 1 & 0 \\ \mathrm{b}(z) & 0 & 0 & 1 \\ -\mathrm{a}(z) & 0 & 0 & -\mathrm{b}(z) \\ 0 & -\mathrm{a}(z) & \mathrm{b}(z) & 0 \end{pmatrix}, \tag{156b}$$

if we write

$$\begin{aligned} \widetilde{\mathsf{H}}_p &= \frac{1}{2n_0}\wp_\perp^2 + \frac{1}{2}n_0\,\mathrm{a}(z)\,r_\perp^2 - \mathrm{b}(z)\,L_z - \mathrm{c}(z), \\ \mathrm{a}(z) &= -2\left(\frac{\alpha_2(z)}{n_0} - \frac{\lambda\!\!\!^{-2}}{2n_0^4}\alpha_2^2(z)\right), \\ \mathrm{b}(z) &= \frac{\lambda\!\!\!^{-}}{2n_0^2}\alpha_2(z), \qquad \mathrm{c}(z) = \left(n_0 + \frac{\lambda\!\!\!^{-2}}{2n_0^2}\alpha_2(z)\right). \end{aligned} \tag{157}$$

Let us now observe that

$$\begin{aligned} \widetilde{\mathsf{T}} &= \begin{pmatrix} 0 & 1 \\ -\mathrm{a}(z) & 0 \end{pmatrix} \otimes \begin{pmatrix} 1 & 0 \\ 0 & 1 \end{pmatrix} + \begin{pmatrix} 1 & 0 \\ 0 & 1 \end{pmatrix} \otimes \begin{pmatrix} 0 & -\mathrm{b}(z) \\ \mathrm{b}(z) & 0 \end{pmatrix} \\ &= (t_F(z) \otimes 1\!\!1_2) + (1\!\!1_2 \otimes t_R(z)), \end{aligned} \tag{158}$$

with the direct product of matrices defined by

$$A \otimes B = \begin{pmatrix} A_{11}B & A_{12}B & \ldots & A_{1n}B \\ A_{21}B & A_{22}B & \ldots & A_{2n}B \\ \vdots & \vdots & \vdots & \vdots \\ A_{n1}B & A_{n2}B & \ldots & A_{nn}B \end{pmatrix}, \tag{159}$$

where A is an $n \times n$ matrix and B is, say, an $m \times m$ matrix. Note that n and m need not be equal. The direct product matrix $A \otimes B$ is an $nm \times nm$ matrix. Further, it should be noted that $(A \otimes B)(C \otimes D) = (AC) \otimes (BD)$, provided A and C have the same dimension and B and D have the same dimension. Hence, if A and C commute and B and D commute, i.e., $AC = CA$ and $BD = DB$, then $(A \otimes B)(C \otimes D) = (C \otimes D)(A \otimes B)$. This shows that in (158) $(1\!\!1_2 \otimes t_R(z))$ commutes with $(t_F(z) \otimes 1\!\!1_2)$.

Let us consider a graded-index (GRIN) fiber tube of length $\Delta z = z - z'$ $(z > z')$ with z' and z referring to the locations of the perpendicular input and output planes, respectively, along the optic axis *i.e.*, z-axis. Let a ray enter the tube at the input plane at $z = z'$ with the ray coordinates $(x(z'), y(z'), \wp_x(z'), \wp_y(z'))$. Let us observe the ray coordinates in the output plane at z to be $(x(z), y(z), \wp_x(z), \wp_y(z))$. The relation between the ray coordinates in the input and output planes is given by the transfer map obtained by integrating Hamilton's equations (156). Integrating (156) we shall write the phase space transfer map as

$$\begin{pmatrix} x(z) \\ y(z) \\ \frac{\wp_x(z)}{n_0} \\ \frac{\wp_y(z)}{n_0} \end{pmatrix} = \widetilde{\mathcal{T}}(z, z') \begin{pmatrix} x(z') \\ y(z') \\ \frac{\wp_x(z')}{n_0} \\ \frac{\wp_y(z')}{n_0} \end{pmatrix}. \tag{160}$$

The transfer matrix $\widetilde{\mathcal{T}}(z, z')$ is obtained following (60). Thus,

$$\widetilde{\mathcal{T}}(z, z') = e^{\widetilde{T}(z, z')}, \tag{161a}$$

$$\begin{aligned} \widetilde{T}(z, z') &= \int_{z'}^{z} dz\, \widetilde{\mathsf{T}}(z) + \frac{1}{2} \int_{z'}^{z} dz_2 \int_{z'}^{z_2} dz_1 \left[\widetilde{\mathsf{T}}(z_2), \widetilde{\mathsf{T}}(z_1)\right] + \cdots \\ &= \left[\left(\int_{z'}^{z} dz\, t_F(z) + \frac{1}{2} \int_{z'}^{z} dz_2 \int_{z'}^{z_2} dz_1 \left[t_F(z_2), t_F(z_1)\right] + \ldots\right) \otimes \mathbb{1}_2\right] \\ &+ \left[\mathbb{1}_2 \otimes \left(\int_{z'}^{z} dz\, t_R(z)\right)\right]. \end{aligned} \tag{161b}$$

In deriving this equation we have used the fact that $(\mathbb{1}_2 \otimes t_R(z))$ commutes with $(t_F(z) \otimes \mathbb{1}_2)$, and $[t_R(z_2), t_R(z_1)] = 0$ for any z_1 and z_2.

Besides the identities in (66), the following two identities are very useful:

$$\begin{aligned} \exp(X)\exp(Y) &= \exp\Big(X + Y + \frac{1}{2}[X, Y] \\ &\quad + \frac{1}{12}[X, [X, Y]] - \frac{1}{12}[Y, [X, Y]] + \cdots\Big), \end{aligned} \tag{162}$$

is the Baker-Campbell-Hausdorff (BCH) formula which helps combine exponentials of matrices, or operators, and

$$\begin{aligned} \exp(X + Y) &= \exp(X)\exp(Y)\exp\left(-\frac{1}{2}[X, Y]\right) \\ &\quad \times \exp\left(\frac{1}{3}[Y, [X, Y]] + \frac{1}{6}[X, [X, Y]]\right)\ldots, \end{aligned} \tag{163}$$

is the Zassenhaus formula which helps split exponentials of matrices, or operators. From (162), or (163), it follows that

$$e^{X+Y} = e^X e^Y, \qquad \text{if } XY = YX. \tag{164}$$

Using this fact, we can write

$$\begin{aligned}
\widetilde{\mathcal{T}}(z, z') &= \exp\Bigg[\Bigg(\int_{z'}^{z} dz\; t_F(z) \\
&\quad + \frac{1}{2}\int_{z'}^{z} dz_2 \int_{z'}^{z_2} dz_1\, [t_F(z_2), t_F(z_1)] + \cdots\Bigg) \otimes 1_2\Bigg] \\
&\quad \times \exp\left[1_2 \otimes \left(\int_{z'}^{z} dz\; t_R(z)\right)\right] \\
&= \exp\Bigg(\int_{z'}^{z} dz\; t_F(z) \\
&\quad + \frac{1}{2}\int_{z'}^{z} dz_2 \int_{z'}^{z_2} dz_1\, [t_F(z_2), t_F(z_1)] + \cdots\Bigg) \\
&\quad \otimes \exp\left(\int_{z'}^{z} dz\; t_R(z)\right).
\end{aligned} \tag{165}$$

In the expression for $\widetilde{\mathcal{T}}(z, z')$ in (165) the second factor can be written down exactly while the calculation of the first factor depends on the exact form of $a(z)$. In general, we can write

$$\widetilde{\mathcal{T}}(z, z') = \begin{pmatrix} \mathrm{A}(z, z') & \mathrm{B}(z, z') \\ \mathrm{C}(z, z') & \mathrm{D}(z, z') \end{pmatrix} \otimes \begin{pmatrix} \cos\theta(z, z') & \sin\theta(z, z') \\ -\sin\theta(z, z') & \cos\theta(z, z') \end{pmatrix}, \tag{166}$$

where

$$\theta(z, z') = -\int_{z'}^{z} dz\; \mathrm{b}(z) = -\frac{\lambda\!\!\!^{-}}{2n_0^2}\int_{z'}^{z} dz\; \alpha_2(z). \tag{167}$$

Equation (166) shows that there is a wavelength-dependent intrinsic continuous twist in the propagation of the ray through the axially symmetric graded-index medium arising from the angular momentum term ($\sim L_z$) in the Hamiltonian (154).

In the coordinate frame rotating with the ray Hamilton's equations become,

$$\frac{d}{dz}\begin{pmatrix} x \\ y \\ \frac{\wp_x}{n_0} \\ \frac{\wp_y}{n_0} \end{pmatrix} = \begin{pmatrix} 0 & 0 & 1 & 0 \\ 0 & 0 & 0 & 1 \\ -\mathrm{a}(z) & 0 & 0 & 0 \\ 0 & -\mathrm{a}(z) & 0 & 0 \end{pmatrix}\begin{pmatrix} x \\ y \\ \frac{\wp_x}{n_0} \\ \frac{\wp_y}{n_0} \end{pmatrix}. \tag{168}$$

In other words,

$$\frac{d\boldsymbol{r}_{\perp}}{dz} = \frac{\wp_{\perp}}{n_0}, \tag{169a}$$

$$\frac{d^2\boldsymbol{r}_{\perp}}{dz^2} = \frac{d}{dz}\left(\frac{\wp_{\perp}}{n_0}\right) = -\,\mathrm{a}(z)\,\boldsymbol{r}_{\perp}. \tag{169b}$$

Thus we have the paraxial ray equation

$$\frac{d^2\boldsymbol{r}_{\perp}}{dz^2} - \left(\frac{2\alpha_2(z)}{n_0} - \frac{\lambda\!\!\!^{-2}}{n_0^4}\alpha_2^2(z)\right)\boldsymbol{r}_{\perp} = 0. \tag{170}$$

Let $\mathrm{A}(z, z')$ and $\mathrm{B}(z, z')$ be two linearly independent solutions of this paraxial ray equation corresponding to the initial conditions

$$\mathrm{A}(z', z') = 1, \quad \left.\frac{d\,\mathrm{A}(z, z')}{dz}\right|_{z=z'} = 0, \tag{171}$$

and

$$\mathrm{B}(z', z') = 0, \quad \left.\frac{d\,\mathrm{B}(z, z')}{dz}\right|_{z=z'} = 1, \tag{172}$$

respectively. Then the general solution of the paraxial ray equation is given by

$$\boldsymbol{r}_{\perp}(z) = \mathrm{A}(z, z')\boldsymbol{r}_{\perp}(z') + \mathrm{B}(z, z')\left(\frac{\wp_{\perp}(z')}{n_0}\right), \tag{173a}$$

$$\begin{aligned} \frac{\wp_{\perp}(z)}{n_0} &= \mathrm{C}(z, z')\boldsymbol{r}_{\perp}(z') + \mathrm{D}(z, z')\left(\frac{\wp_{\perp}(z')}{n_0}\right), \\ &\text{with} \quad \mathrm{C}(z, z') = \frac{d\,\mathrm{A}(z, z')}{dz}, \\ &\mathrm{D}(z, z') = \frac{d\,\mathrm{B}(z, z')}{dz}. \end{aligned} \tag{173b}$$

Equation (173b) follows from the relation (169a). The matrix elements $\mathrm{A}(z, z')$, $\mathrm{B}(z, z')$, $\mathrm{C}(z, z')$, and $\mathrm{D}(z, z')$, in (166) are the same as the coefficients in (173). Further, it may be noted that since $t_F(z)$ and the commutators $[t_F(z_2), t_F(z_1)]$ in the exponent in the first exponential factor in (165) are tracelss matrices,

$$\det\begin{pmatrix} \mathrm{A}(z, z') & \mathrm{B}(z, z') \\ \mathrm{C}(z, z') & \mathrm{D}(z, z') \end{pmatrix} = 1, \tag{174}$$

the condition necessary for preserving the symplectic structure of phase space under the transformation (173).

5.2.2 GRIN lens: Image rotation

For an axially symmetric GRIN lens the refractive index profile is given by

$$n(\boldsymbol{r}) = n_0\left(1 - \frac{1}{2}\alpha^2 r_\perp^2\right), \tag{175}$$

where α is a constant. Thus, for the GRIN lens we shall take

$$\alpha_2(z) = -\frac{1}{2}n_0\alpha^2. \tag{176}$$

Then, the paraxial ray equation (170) becomes

$$\frac{d^2\boldsymbol{r}_\perp}{dz^2} + \widetilde{\alpha}^2\boldsymbol{r}_\perp = 0, \qquad \widetilde{\alpha} = \alpha\sqrt{1 + \frac{\bar{\lambda}^2}{4n_0^2}} \approx \alpha\left(1 + \frac{\bar{\lambda}^2}{8n_0^2}\right). \tag{177}$$

The two linearly independent solutions of (177), satisfying the prescribed initial conditions (171) and (172), respectively, are

$$\begin{aligned} \mathrm{A}(z, z') &= \cos((z - z')\widetilde{\alpha}), \\ \mathrm{B}(z, z') &= \tfrac{1}{\widetilde{\alpha}}\sin((z - z')\widetilde{\alpha}). \end{aligned} \tag{178}$$

Hence, from (173b), we have,

$$\begin{aligned} \mathrm{C}(z, z') &= -\widetilde{\alpha}\sin((z - z')\widetilde{\alpha}), \\ \mathrm{D}(z, z') &= \cos((z - z')\widetilde{\alpha}). \end{aligned} \tag{179}$$

Note that $\mathrm{A}(z, z')$, $\mathrm{B}(z, z')$, $\mathrm{C}(z, z')$ and $\mathrm{D}(z, z')$ satisfy (174).

Let a GRIN lens have its left end, or the input end, at $z = z'$ and the right end, or the output end, at $z > z'$. If $z - z' = d$, the length of the lens, then the transfer matrix for the lens becomes

$$\widetilde{\mathcal{T}}_L(z, z') = \begin{pmatrix} \cos(\widetilde{\alpha}d) & \frac{1}{\widetilde{\alpha}}\sin(\widetilde{\alpha}d) \\ -\widetilde{\alpha}\sin(\widetilde{\alpha}d) & \cos(\widetilde{\alpha}d) \end{pmatrix} \otimes \begin{pmatrix} \cos\theta_L & \sin\theta_L \\ -\sin\theta_L & \cos\theta_L \end{pmatrix}, \tag{180}$$

where

$$\theta_L = -\frac{\bar{\lambda}}{2n_0^2}\int_{z'}^{z} dz\ \alpha_2(z) = \frac{\bar{\lambda}\alpha^2 d}{4n_0}. \tag{181}$$

A simple lens of focal length f with drift spaces of lengths D on both sides has the transfer matrix

$$\begin{aligned}\mathcal{T}_{SL} &= \begin{pmatrix}1 & D\\ 0 & 1\end{pmatrix}\begin{pmatrix}1 & 0\\ -\frac{1}{f} & 1\end{pmatrix}\begin{pmatrix}1 & D\\ 0 & 1\end{pmatrix}\\ &= \begin{pmatrix}1-\frac{D}{f} & 2D-\frac{D^2}{f}\\ -\frac{1}{f} & 1-\frac{D}{f}\end{pmatrix}.\end{aligned} \tag{182}$$

Comparing $\widetilde{\mathcal{T}}_L$ with T_{SL} it is seen that the GRIN lens is equivalent to a simple lens with focal length

$$f = \frac{1}{\widetilde{\alpha}\sin(\widetilde{\alpha}d)}, \tag{183}$$

and drift spaces of length

$$D = \frac{1}{\widetilde{\alpha}}\tan\left(\frac{1}{2}\widetilde{\alpha}d\right), \tag{184}$$

on both sides of the lens (Shou et al., 2011; Tsai & Chu, 2013). Further, it should be noted that there is a wavelength-dependent image rotation by an angle θ_L as given in (181) (Khan, 2010, 2014a, 2016c, 2017c,e). As seen from (180) a half-pitch GRIN lens ($\widetilde{\alpha}d \approx \alpha d = \pi$) gives at the output plane at z an inverted image of an object at the input plane at z'. A full-pitch GRIN lens ($\widetilde{\alpha}d \approx \alpha d = 2\pi$) images from z' to z without inversion. For a typical full-pitch GRIN lens with parameters $\alpha = 0.6\,\mathrm{mm}^{-1}$, $\alpha d = 2\pi$, $n_0 = 1.6$, and $\lambda = 600\,\mathrm{nm}$, we find that $\theta_L \approx 6 \times 10^{-5}$! This image rotation is theoretically analogous to the image rotation in an axially symmetric magnetic lens arising from the angular momentum term in the beam optical Hamiltonian (Hawkes & Kasper, 2017a, 2017b; Jagannathan & Khan, 1996, 2019; Khan, 1997).

5.2.3 Aberrations

If the beam propagating in the system is not paraxial, but can be described as quasiparaxial, then paraxial approximation of the wave optical Hamiltonian will not be appropriate. We have to consider in the Hamiltonian terms beyond quadratic in $\boldsymbol{r}_\perp$ and $\widehat{\wp}_\perp$. Thus, for a quasiparaxial light beam propagating in a graded-index medium the Maxwell vector wave optical Hamiltonian is $\widehat{\mathsf{H}}$, given by (144), in which terms up to fourth power in $\boldsymbol{r}_\perp$ and $\widehat{\wp}_\perp$ have been retained. Dropping all the matrices from the

Maxwell vector wave optical Hamiltonian $\widehat{\mathcal{H}}$ we obtain the scalar wave optical Hamiltonian

$$\begin{aligned}\widehat{\mathrm{H}} = & -n_0\left(1 + \frac{\lambda\!\!\!^{-2}}{2n_0^3}\alpha_2(z)\right) + \frac{1}{2n_0}\widehat{\wp}_\perp^2 - \alpha_2(z)\left(1 - \frac{\lambda\!\!\!^{-2}}{2n_0^3}\alpha_2(z)\right)r_\perp^2 \\ & - \frac{\lambda\!\!\!^{-}}{2n_0^2}\alpha_2(z)\widehat{L}_z + \frac{1}{8n_0^3}\widehat{\wp}_\perp^4 - \frac{\alpha_2(z)}{4n_0^2}\left(r_\perp^2\widehat{\wp}_\perp^2 + \widehat{\wp}_\perp^2 r_\perp^2\right) \\ & - \alpha_4(z)\left(1 - \frac{\lambda\!\!\!^{-2}}{2n_0^3}\alpha_2(z)\right)r_\perp^4 + \frac{\lambda\!\!\!^{-}}{4n_0^3}\alpha_2^2(z)\left(r_\perp^2\widehat{L}_z + \widehat{L}_z r_\perp^2\right).\end{aligned} \tag{185}$$

Let us look at the *mathematical* ray picture as in the previous Section. The corresponding ray optical Hamiltonian is obtained from $\widehat{\mathrm{H}}$ using the replacement $(\boldsymbol{r}_\perp, \widehat{\wp}_\perp) \longrightarrow (\boldsymbol{r}_\perp, \wp_\perp)$. Thus, we obtain the ray optical Hamiltonian:

$$\begin{aligned}\widetilde{\mathrm{H}} = & -n_0\left(1 + \frac{\lambda\!\!\!^{-2}}{2n_0^3}\alpha_2(z)\right) + \frac{1}{2n_0}\wp_\perp^2 - \alpha_2(z)\left(1 - \frac{\lambda\!\!\!^{-2}}{2n_0^3}\alpha_2(z)\right)r_\perp^2 \\ & - \frac{\lambda\!\!\!^{-}}{2n_0^2}\alpha_2(z)L_z + \frac{1}{8n_0^3}\wp_\perp^4 - \frac{\alpha_2(z)}{4n_0^2}\left(r_\perp^2\wp_\perp^2 + \wp_\perp^2 r_\perp^2\right) \\ & - \alpha_4(z)\left(1 - \frac{\lambda\!\!\!^{-2}}{2n_0^3}\alpha_2(z)\right)r_\perp^4 + \frac{\lambda\!\!\!^{-}}{4n_0^3}\alpha_2^2(z)\left(r_\perp^2 L_z + L_z r_\perp^2\right).\end{aligned} \tag{186}$$

Now, Hamilton's equation for x is

$$\begin{aligned}\frac{dx}{dz} & = :-\widetilde{\mathrm{H}}:\, x \\ & = -\frac{\lambda\!\!\!^{-}\alpha_2(z)}{2n_0^2}y + \frac{1}{n_0}\wp_x \\ & \quad + \frac{1}{2n_0^3}\wp_x\wp_\perp^2 - \frac{\alpha_2(z)}{n_0^2}\wp_x r_\perp^2 - \frac{\lambda\!\!\!^{-}\alpha_2^2(z)}{2n_0^3}y r_\perp^2.\end{aligned} \tag{187}$$

This result shows that when the Hamiltonian contains terms of fourth order in $(\boldsymbol{r}_\perp, \wp_\perp)$ the z-evolution of $(\boldsymbol{r}_\perp, \wp_\perp)$ does not depend linearly on $(\boldsymbol{r}_\perp, \wp_\perp)$, but depends on terms of third order in $(\boldsymbol{r}_\perp, \wp_\perp)$ besides the linear terms. Due to this nonlinearity the imaging is not point-to-point as in the paraxial situation. The resulting image is said to be affected by third order aberrations.

Since the phase space map is not linear for the system with the Hamiltonian containing nonparaxial terms, it is not possible to describe it using a transfer matrix. To understand the phase space map leading to image aberrations let us look at the scalar wave optical transfer operator

$$\widehat{\mathcal{T}}(z, z') = \exp\left(-\frac{i}{\lambda\!\!\!^{-}}\widehat{T}(z, z')\right), \tag{188}$$

where $\widehat{T}(z, z')$ is to be calculated using the Magnus formula (60). To proceed further, we shall be adopting techniques from time-dependent perturbation theory in quantum mechanics. The z-evolution of the scalar beam wave function $\Psi(\boldsymbol{r}_\perp; z)$ is given by

$$i\lambda\!\!\!^{-}\frac{\partial\Psi(\boldsymbol{r}_\perp; z)}{\partial z} = \widehat{\mathsf{H}}\Psi(\boldsymbol{r}_\perp; z). \tag{189}$$

Let us write

$$\widehat{\mathsf{H}} = \widehat{\mathsf{H}}_p + \widehat{\mathsf{H}}', \tag{190}$$

where

$$\begin{aligned} \widehat{\mathsf{H}}_p &= -n_0\left(1 + \frac{\lambda\!\!\!^{-2}}{2n_0^3}\alpha_2(z)\right) + \frac{1}{2n_0}\widehat{\wp}_\perp^2 \\ &\quad - \alpha_2(z)\left(1 - \frac{\lambda\!\!\!^{-2}}{2n_0^3}\alpha_2(z)\right)r_\perp^2 - \frac{\lambda\!\!\!^{-}}{2n_0^2}\alpha_2(z)\widehat{L}_z, \end{aligned} \tag{191}$$

is the paraxial part and

$$\begin{aligned} \widehat{\mathsf{H}}' &= \frac{1}{8n_0^3}\widehat{\wp}_\perp^4 - \frac{\alpha_2(z)}{4n_0^2}\left(r_\perp^2\widehat{\wp}_\perp^2 + \widehat{\wp}_\perp^2 r_\perp^2\right) \\ &\quad - \alpha_4(z)\left(1 - \frac{\lambda\!\!\!^{-2}}{2n_0^3}\alpha_2(z)\right)r_\perp^4 + \frac{\lambda\!\!\!^{-}}{4n_0^3}\alpha_2^2(z)\left(r_\perp^2\widehat{L}_z + \widehat{L}_z r_\perp^2\right), \end{aligned} \tag{192}$$

is the aberration, or perturbation, part. Adopting the interaction picture used in the time-dependent perturbation theory in quantum mechanics, let us define

$$\begin{aligned} \Psi_{\mathrm{I}}(z) &= \left(\widehat{\mathcal{T}}_p(z, z')\right)^{-1}\Psi(z), \\ \widehat{\mathcal{T}}_p(z, z') &= \mathbb{P}\left[\exp\left(-\frac{i}{\lambda\!\!\!^{-}}\int_{z'}^{z} dz\,\widehat{\mathsf{H}}_p(z)\right)\right], \end{aligned} \tag{193}$$

where we are suppressing $\boldsymbol{r}_\perp$ dependence of Ψ. Note that, since $\widehat{\mathcal{T}}_p(z', z') = I$,

$$\Psi_{\mathrm{I}}(z') = \left(\widehat{\mathcal{T}}_p(z', z')\right)^{-1}\Psi(z') = \Psi(z'). \tag{194}$$

Defining

$$\widehat{\mathsf{H}}_{\mathrm{I}}' = \left(\widehat{\mathcal{T}}_p(z, z')\right)^{-1}\widehat{\mathsf{H}}'\widehat{\mathcal{T}}_p(z, z'), \tag{195}$$

one can show that

$$i\lambda\!\!\!^{-}\frac{\partial\Psi_{\mathrm{I}}}{\partial z} = \widehat{\mathsf{H}}_{\mathrm{I}}'\Psi_{\mathrm{I}}. \tag{196}$$

Integrating this equation we get

$$\begin{aligned} \Psi_{\mathrm{I}}(z) &= \widehat{\mathcal{T}}_{\mathrm{I}}(z, z')\Psi_{\mathrm{I}}(z') = \widehat{\mathcal{T}}_{\mathrm{I}}(z, z')\Psi(z'), \\ \widehat{\mathcal{T}}_{\mathrm{I}}(z, z') &= \mathbb{P}\left[\exp\left(-\frac{i}{\lambda\!\!\!^{-}}\int_{z'}^{z} dz\, \widehat{\mathsf{H}}_{\mathrm{I}}'(z)\right)\right]. \end{aligned} \tag{197}$$

Then, from (193) and (197) we get

$$\Psi(z) = \widehat{\mathcal{T}}_{p}(z, z')\widehat{\mathcal{T}}_{\mathrm{I}}(z, z')\Psi(z'), \tag{198}$$

describing the complete transfer map.

Using the Magnus formula, and dropping all terms of degree higher than four in $(\boldsymbol{r}_{\perp}, \widehat{\wp}_{\perp})$, we have

$$\begin{aligned} \widehat{\mathcal{T}}_{\mathrm{I}}(z, z') &= \exp\left(-\frac{i}{\lambda\!\!\!^{-}}\widehat{T}_{\mathrm{I}}(z, z')\right) \\ &\approx \exp\left(-\frac{i}{\lambda\!\!\!^{-}}\int_{z'}^{z} dz\, \widehat{\mathsf{H}}_{\mathrm{I}}'(z)\right) \\ &= \exp\left(-\frac{i}{\lambda\!\!\!^{-}}\int_{z'}^{z} dz\, (\mathcal{T}_{p}(z, z'))^{-1}\widehat{\mathsf{H}}'\mathcal{T}_{p}(z, z')\right), \end{aligned} \tag{199}$$

with

$$\begin{aligned} \widehat{T}_{\mathrm{I}}(z, z') = \{ & C(z, z')\widehat{\wp}_{\perp}^{4} \\ & + K(z, z')\left[\widehat{\wp}_{\perp}^{2}, (\widehat{\wp}_{\perp}\cdot\boldsymbol{r}_{\perp} + \boldsymbol{r}_{\perp}\cdot\widehat{\wp}_{\perp})\right]_{+} \\ & + k(z, z')\widehat{\wp}_{\perp}^{2}\widehat{L}_{z} \\ & + A(z, z')(\widehat{\wp}_{\perp}\cdot\boldsymbol{r}_{\perp} + \boldsymbol{r}_{\perp}\cdot\widehat{\wp}_{\perp})^{2} \\ & + a(z, z')(\widehat{\wp}_{\perp}\cdot\boldsymbol{r}_{\perp} + \boldsymbol{r}_{\perp}\cdot\widehat{\wp}_{\perp})\widehat{L}_{z} \\ & + F(z, z')(\widehat{\wp}_{\perp}^{2}r_{\perp}^{2} + r_{\perp}^{2}\widehat{\wp}_{\perp}^{2}) \\ & + D(z, z')\left[r_{\perp}^{2}, (\widehat{\wp}_{\perp}\cdot\boldsymbol{r}_{\perp} + \boldsymbol{r}_{\perp}\cdot\widehat{\wp}_{\perp})\right]_{+} \\ & + d(z, z')r_{\perp}^{2}\widehat{L}_{z} \\ & + E(z, z')r_{\perp}^{4}\}. \end{aligned} \tag{200}$$

The nine aberration coefficients occurring in the above expression for $\widehat{\mathcal{T}}_i(z, z')$ are given in terms of $\mathrm{A}(z, z')$, $\mathrm{B}(z, z')$, $\mathrm{C}(z, z')$, and $\mathrm{D}(z, z')$, the matrix elements of the paraxial transfer matrix $\mathcal{T}(z, z')$ in (166), or the solutions for the ray coordinates $\boldsymbol{r}_\perp$ and $\wp_\perp / n_0$ obtained from the paraxial ray equation (170), as follows:

$$
\begin{aligned}
C(z, z') &= \int_{z'}^{z} dz \left\{ \frac{1}{8n_0^3}\mathrm{D}^4 - \frac{\alpha_2(z)}{2n_0^2}\mathrm{B}^2\mathrm{D}^2 \right. \\
&\quad \left. - \alpha_4(z)\left(1 - \frac{\lambda\!\!\!^{-2}}{2n_0^3}\alpha_2(z)\right)\mathrm{B}^4 \right\} \\
&\qquad\qquad \text{(Spherical Aberration)}, \\
K(z, z') &= \int_{z'}^{z} dz \left\{ \frac{1}{8n_0^3}\mathrm{C}\mathrm{D}^3 - \frac{\alpha_2(z)}{4n_0^2}\mathrm{B}\mathrm{D}(\mathrm{A}\mathrm{D} + \mathrm{B}\mathrm{C}) \right. \\
&\quad \left. - \alpha_4(z)\left(1 - \frac{\lambda\!\!\!^{-2}}{2n_0^3}\alpha_2(z)\right)\mathrm{A}\mathrm{B}^3 \right\} \\
&\qquad\qquad \text{(Coma)}, \\
k(z, z') &= \frac{\lambda\!\!\!^{-}}{2n_0^3}\int_{z'}^{z} dz\, \alpha_2^2(z)\mathrm{B}^2, \qquad \text{(Anisotropic Coma)} \\
A(z, z') &= \int_{z'}^{z} dz \left\{ \frac{1}{8n_0^3}\mathrm{C}^2\mathrm{D}^2 - \frac{\alpha_2(z)}{2n_0^2}\mathrm{A}\mathrm{B}\mathrm{C}\mathrm{D} \right. \\
&\quad \left. - \alpha_4(z)\left(1 - \frac{\lambda\!\!\!^{-2}}{2n_0^3}\alpha_2(z)\right)\mathrm{A}^2\mathrm{B}^2 \right\} \\
&\qquad\qquad \text{(Astigmatism)}, \\
a(z, z') &= \frac{\lambda\!\!\!^{-}}{2n_0^3}\int_{z'}^{z} dz\, \alpha_2^2(z)\mathrm{A}\mathrm{B} \quad \text{(Anisotropic Astigmatism)}, \\
F(z, z') &= \int_{z'}^{z} dz \left\{ \frac{1}{8n_0^3}\mathrm{C}^2\mathrm{D}^2 - \frac{\alpha_2(z)}{4n_0^2}(\mathrm{A}^2\mathrm{D}^2 + \mathrm{B}^2\mathrm{C}^2) \right. \\
&\quad \left. - \alpha_4(z)\left(1 - \frac{\lambda\!\!\!^{-2}}{2n_0^3}\alpha_2(z)\right)\mathrm{A}^2\mathrm{B}^2 \right\} \\
&\qquad\qquad \text{(Field Curvature)}, \\
D(z, z') &= \int_{z'}^{z} dz \left\{ \frac{1}{8n_0^3}\mathrm{C}^3\,\mathrm{D} - \frac{\alpha_2(z)}{4n_0^2}\mathrm{A}\mathrm{C}(\mathrm{A}\mathrm{D} + \mathrm{B}\mathrm{C}) \right. \\
&\quad \left. - \alpha_4(z)\left(1 - \frac{\lambda\!\!\!^{-2}}{2n_0^3}\alpha_2(z)\right)\mathrm{A}^3\mathrm{C} \right\} \\
&\qquad\qquad \text{(Distortion)}, \\
d(z, z') &= \frac{\lambda\!\!\!^{-}}{2n_0^3}\int_{z'}^{z} dz\, \alpha_2^2(z)\mathrm{A}^2 \quad \text{(Anisotropic Distortion)},
\end{aligned}
\tag{201}
$$

and

$$
\begin{aligned}
E(z, z') = \int_{z'}^{z} dz \Big\{ & \tfrac{1}{8n_0^3}\mathrm{C}^4 - \tfrac{\alpha_2(z)}{2n_0^2}\mathrm{A}^2\mathrm{C}^2 \\
& - \alpha_4(z)\Big(1 - \tfrac{\lambdabar^2}{2n_0^3}\alpha_2(z)\Big)\mathrm{A}^4\Big\} \quad \text{(Pocus).}
\end{aligned}
\tag{202}
$$

The name Pocus is due to Kurt Bernardo Wolf (see page 137 in Dragt et al., 1986). Due to the axial symmetry of the graded index system considered, the nine third-order aberration terms in the Hamiltonian $\widehat{\mathsf{H}}$ in (185) are composed of four elements $\widehat{\wp}_\perp^2$, $r_\perp^2$, $(\widehat{\wp}_\perp \cdot \boldsymbol{r}_\perp + \boldsymbol{r}_\perp \cdot \widehat{\wp}_\perp)$ and $\widehat{L}_z$. This is also true for the higher order Hamiltonians. Combinatorially, one would expect a tenth term namely, $\widehat{L}_z^2$. This term is not listed separately as it can be decomposed into two terms already present in the list of nine aberrations. This decomposition is as follows:

$$
\widehat{L}_z^2 = \frac{1}{2}\left(\widehat{\wp}_\perp^2 r_\perp^2 + r_\perp^2 \widehat{\wp}_\perp^2\right) - \frac{1}{4}\left(\widehat{\wp}_\perp \cdot \boldsymbol{r}_\perp + \boldsymbol{r}_\perp \cdot \widehat{\wp}_\perp\right)^2 + \lambdabar^2. \tag{203}
$$

There are only nine third-order aberrations permitted by axial symmetry. In the Maxwell vector wave optical Hamiltonian $\widehat{\mathcal{H}}$ in (144) the $\widehat{\mathcal{H}}_{0,(4)}$ part is the familiar one from the traditional formalism and is responsible for six aberrations, namely, spherical aberration, coma, astigmatism, field curvature, distortion, and pocus. This part is accompanied by the scalar part, $\mathcal{H}_{0,(4)}^{(\lambdabar)}$, and the matrix part, $\mathcal{H}^{(\lambdabar,\sigma)}$. The terms in the scalar part $\mathsf{H}_{0,(4)}^{(\lambdabar)}$ modify the aforementioned six aberrations by wavelength-dependent corrections. Furthermore, these terms give rise to the remaining three aberrations permitted by the axial symmetry, namely, anisotropic coma, anisotropic astigmatism, and anisotropic distortion. It is seen that all these three aberrations are pure wavelength-dependent and all are anisotropic.

Thus, we have explicitly obtained all the nine third-order aberration coefficients permitted by axial symmetry. Of these, six are present in the Lie algebraic formalism of ray optics (Dragt & Forest, 1986; Dragt et al., 1986). In our approach these six get modified by wavelength-dependent corrections. The remaining three, namely, anisotropic coma, anisotropic astigmatism, and anisotropic distortion, are extra and are purely wavelength-dependent aberrations. It is worth noting that these three aberrations are absent also in the non-traditional formalism of the Helmholtz scalar wave optics (Khan, 2005b, 2016b, 2018a, 2018c; Khan et al., 2002).

6. Transition from the Helmholtz scalar wave optics to the Maxwell vector wave optics

We have derived the exact Maxwell vector wave optical Hamiltonian without any assumptions on the form of the spatially varying refractive index $n(\boldsymbol{r})$. In this Section, we focus on the procedure for transition from the Helmholtz scalar wave optics to the Maxwell vector wave optics. Such a procedure was originally derived by analyzing the relativistic symmetry of the Maxwell equations and is now known as the Mukunda-Simon-Sudarshan matrix substitution rule, or simply the MSS substitution rule, or just the MSS rule (Mukunda et al., 1983, 1985a, 1985b; Simon et al., 1986, 1987; Sudarshan et al., 1983). The same rule emerges from the matrix formalism of the Maxwell vector wave optics (Khan, 2016c, 2017c, 2017e; Khan & Jagannathan, 2024c). First we shall derive the MSS rule from the matrix formalism of the Maxwell vector wave optics. Then we shall present a simpler to use matrix differential operator form of the MSS rule (Khan, 2023b).

6.1 Mukunda-Simon-Sudarshan matrix substitution rule

Let us consider a monochromatic paraxial beam moving in the forward z-direction through an inhomogeneous medium with a srefractive index $n(\boldsymbol{r}_\perp; z)$. From (71) we can write

$$i\lambda\!\!\!^{-}\frac{\partial\overline{\Phi}}{\partial z} = \widehat{\mathcal{H}}_p\overline{\Phi}, \qquad \widehat{\mathcal{H}}_p \approx \mathcal{B}\left(-n(\boldsymbol{r}_\perp; z) + \frac{1}{2n_c}\widehat{\wp}_\perp^2\right), \tag{204}$$

where $\Phi = \overline{\Phi}e^{-i\omega t}$ represents the paraxial beam and $\widehat{\mathcal{H}}_p$ is the Maxwell vector wave optical paraxial Hamiltonian. Dropping the matrix $\mathcal{B}$ from $\widehat{\mathcal{H}}_p$ we get

$$i\lambda\!\!\!^{-}\frac{\partial\overline{\Phi}}{\partial z} = \widehat{\mathsf{H}}_p\overline{\Phi}, \qquad \widehat{\mathsf{H}}_p \approx \left(-n(\boldsymbol{r}_\perp; z) + \frac{1}{2n_c}\widehat{\wp}_\perp^2\right), \tag{205}$$

the Helmholtz scalar wave optical z-evolution equation for $\overline{\Phi}$ in the paraxial approximation. In scalar wave optics we would have normally replaced $\overline{\Phi}$ by a scalar wave function. But, now, let us proceed without replacing $\overline{\Phi}$. Integrating (205), and using the Magnus formula (60), we obtain

$$\begin{aligned}\overline{\Phi}(z) &= \mathbb{P}\left[\exp\left\{-\frac{i}{\lambda\!\!\!^{-}}\int_{z'}^{z} dz\, \widehat{\mathsf{H}}_p(\boldsymbol{r}_\perp, \widehat{\wp}_\perp, z)\right\}\right]\overline{\Phi}(z') \\ &= \exp\left(-\frac{i}{\lambda\!\!\!^{-}}\widehat{T}_p(\boldsymbol{r}_\perp, \widehat{\wp}_\perp; z, z')\right)\overline{\Phi}(z'),\end{aligned} \tag{206}$$

for $z \geq z'$. Since

$$\overline{\Phi} = e^{i\omega t}\mathcal{T}^{\dagger}\mathcal{F}, \tag{207}$$

and $\mathcal{T}$ is a numerical matrix commuting with $\widehat{T}_p$, (206) implies

$$\mathcal{F}(z) = \exp\left(-\frac{i}{\lambdabar}\widehat{T}_p\left(\boldsymbol{r}_\perp, \widehat{\wp}_\perp; z, z'\right)\right)\mathcal{F}(z'). \tag{208}$$

For future notational convenience, let us write

$$\exp\left(-\frac{i}{\lambdabar}\widehat{T}_p\left(\boldsymbol{r}_\perp, \widehat{\wp}_\perp; z, z'\right)\right) = \widehat{\mathcal{G}}\left(\boldsymbol{r}_\perp, \widehat{\wp}_\perp; z, z'\right). \tag{209}$$

Then, we have

$$\begin{pmatrix} \sqrt{\epsilon}\boldsymbol{E}(z) \\ \frac{1}{\sqrt{\mu}}\boldsymbol{B}(z) \end{pmatrix} = \widehat{\mathcal{G}}\left(\boldsymbol{r}_\perp, \widehat{\wp}_\perp; z, z'\right)\begin{pmatrix} \sqrt{\epsilon}\boldsymbol{E}(z') \\ \frac{1}{\sqrt{\mu}}\boldsymbol{B}(z') \end{pmatrix}. \tag{210}$$

For a homogeneous medium with constant ϵ and μ, or for an inhomogeneous medium with $1/\sqrt{\epsilon}$ varying slowly in the xy-plane such that $1/\sqrt{\epsilon}$ commutes with $\widehat{\wp}_\perp$, we can write

$$\mathbb{F}(z) = \widehat{\mathcal{G}}\left(\boldsymbol{r}_\perp, \widehat{\wp}_\perp; z, z'\right)\mathbb{F}(z'), \qquad \mathbb{F}(z) = \begin{pmatrix} \boldsymbol{E}(\boldsymbol{r}_\perp; z) \\ v\boldsymbol{B}(\boldsymbol{r}_\perp; z) \end{pmatrix}, \tag{211}$$

where $v = 1/\sqrt{\epsilon\mu}$.

Let us now consider the propagation of a monochromatic paraxial beam in a homogeneous medium of constant refractive index n_c. If we define

$$\overline{\Phi}_c = \widehat{T}_c\overline{\Phi}, \tag{212}$$

where $\widehat{T}_c$ is as given in (105), we have, from (106),

$$\begin{aligned} i\lambdabar\frac{\partial\overline{\Phi}_c}{\partial z} &= -\mathcal{B}\widehat{\wp}_z\overline{\Phi}_c, \\ -\widehat{\wp}_z &= -\sqrt{n_c^2 - \widehat{\wp}_\perp^2} \approx -n_c + \frac{1}{2n_c}\widehat{\wp}_\perp^2. \end{aligned} \tag{213}$$

Explicitly writing,

$$\overline{\Phi}_c = \frac{1}{2\sqrt{2\widehat{\wp}_z\left(n_c + \widehat{\wp}_z\right)}}\begin{pmatrix} \left(n_c + \widehat{\wp}_z\right)\left(-F_x^+ + iF_y^+\right) - \widehat{\wp}_- F_z^+ \\ \left(n_c + \widehat{\wp}_z\right)F_z^+ - \widehat{\wp}_-\left(F_x^+ + iF_y^+\right) \\ \left(n_c + \widehat{\wp}_z\right)\left(-F_x^- - iF_y^-\right) - \widehat{\wp}_+ F_z^- \\ \left(n_c + \widehat{\wp}_z\right)F_z^- - \widehat{\wp}_+\left(F_x^- - iF_y^-\right) \\ \left(n_c + \widehat{\wp}_z\right)F_z^+ - \widehat{\wp}_+\left(-F_x^+ + iF_y^+\right) \\ \left(n_c + \widehat{\wp}_z\right)\left(F_x^+ + iF_y^+\right) - \widehat{\wp}_+ F_z^+ \\ \left(n_c + \widehat{\wp}_z\right)F_z^- - \widehat{\wp}_-\left(-F_x^- - iF_y^-\right) \\ \left(n_c + \widehat{\wp}_z\right)\left(F_x^- - iF_y^-\right) - \widehat{\wp}_- F_z^- \end{pmatrix}. \tag{214}$$

A plane wave moving in the positive z-direction close to the z-axis can be represented by

$$\begin{aligned} &\overline{\Phi}_c\left(\boldsymbol{r}_\perp;\, z\right) = \overline{\Phi}_0 \exp\left\{in_c\left(\boldsymbol{k}_\perp\cdot\boldsymbol{r}_\perp + k_z z\right)\right\}, \\ &k_z > 0, \quad k_x^2, \quad k_y^2 \ll k^2, \end{aligned} \tag{215}$$

where the components of $\overline{\Phi}_0$ are constants, and $n_c\boldsymbol{k}$ is the wave vector in the medium. The plane wave in (215) has to satisfy (213). This leads to $-n_c k_z/k = -\sqrt{n_c^2 - \left(n_c^2 k_\perp^2/k^2\right)} \approx -n_c\left(1 - \left(k_\perp^2/2k^2\right)\right)$ for the upper four components of $\overline{\Phi}_c$ and $-n_c k_z/k = \sqrt{n_c^2 - \left(n_c^2 k_\perp^2/k^2\right)} \approx n_c\left(1 - \left(k_\perp^2/2k^2\right)\right)$ for the lower four components of $\overline{\Phi}_c$. This implies that, since $k_z > 0$ and $k_\perp^2 \ll k^2$, all the four lower components of $\overline{\Phi}_c$ associated with a plane wave moving in the $+z$-direction in the medium must vanish. This can be easily verified from (214) for the example of a plane wave moving in the $+z$-direction and polarized in the x-direction.

A paraxial beam with a finite transverse extent in the xy-plane propagating predominantly in the forward z-direction can be written as a superposition of plane waves of the type in (215). Explicitly, for any such beam satisfying the paraxial condition k_x^2, $k_y^2 \ll k^2$ we can write

$$\begin{aligned} \boldsymbol{F}^\pm &= \exp(ikz)\int_{-\infty}^{\infty}\int_{-\infty}^{\infty} dk_x\, dk_y\, \tilde{\boldsymbol{F}}^\pm\left(k_x,\, k_y\right) \\ &\quad\times \exp\left\{i\left[k_x x + k_y y - z\left(\frac{k_x^2 + k_y^2}{2k}\right)\right]\right\}. \end{aligned} \tag{216}$$

In other words,

$$
\begin{aligned}
\overline{\Phi}_c(\boldsymbol{r}_\perp; z) = {} & \frac{1}{2}\exp\{ikz\}\int_{-\infty}^{\infty}\int_{-\infty}^{\infty} dk_x\, dk_y \frac{1}{\sqrt{2\wp_z(n_c+\wp_z)}} \\
& \times \begin{pmatrix}
(n_c+\wp_z)(-\tilde{F}_x^+ + i\tilde{F}_y^+) - \wp_-\tilde{F}_z^+ \\
(n_c+\wp_z)\tilde{F}_z^+ - \wp_-(\tilde{F}_x^+ + i\tilde{F}_y^+) \\
(n_c+\wp_z)(-\tilde{F}_x^- - i\tilde{F}_y^-) - \wp_+\tilde{F}_z^- \\
(n_c+\wp_z)\tilde{F}_z^- - \wp_+(\tilde{F}_x^- - i\tilde{F}_y^-) \\
(n_c+\wp_z)\tilde{F}_z^+ - \wp_+(-\tilde{F}_x^+ + i\tilde{F}_y^+) \\
(n_c+\wp_z)(\tilde{F}_x^+ + i\tilde{F}_y^+) - \wp_+\tilde{F}_z^+ \\
(n_c+\wp_z)\tilde{F}_z^- - \wp_-(-\tilde{F}_x^- - i\tilde{F}_y^-) \\
(n_c+\wp_z)(\tilde{F}_x^- - i\tilde{F}_y^-) - \wp_-\tilde{F}_z^-
\end{pmatrix} \\
& \times \exp\left\{i\left[k_x x + k_y y - z\left(\frac{k_x^2+k_y^2}{2k}\right)\right]\right\},
\end{aligned}
\tag{217}
$$

where $\wp_\pm = n_c(k_x \pm ik_y)/k$, $\wp_z \approx n_c(1-(k_\perp^2/2k^2))$, and the Fourier coefficients $\tilde{\boldsymbol{F}}^\pm(k_x, k_y)$ are nonzero only within a small range of $k_x, k_y \ll k$. The vanishing of all the four lower components of $\overline{\Phi}_c(\boldsymbol{r}_\perp, z)$ for a paraxial beam propagating in the forward z-direction in a homogeneous medium implies a set of relationships among the components of the $\boldsymbol{E}$ and $\boldsymbol{B}$ fields of the beam and this leads to interesting consequences. These constraints imply that in the region of free space,

$$
\begin{aligned}
E_z &= \frac{i\bar{\lambda}}{(n_c+\widehat{\wp}_z)}((\nabla_\perp\cdot\boldsymbol{E}_\perp) + v(\nabla\times\boldsymbol{B})_z), \\
B_z &= \frac{i\bar{\lambda}}{(n_c+\widehat{\wp}_z)}\left((\nabla_\perp\cdot\boldsymbol{B}_\perp) - \frac{1}{v}(\nabla\times\boldsymbol{E})_z\right), \\
B_x &= -\frac{1}{v}E_y - \frac{i\bar{\lambda}}{(n_c+\widehat{\wp}_z)}\left(\frac{1}{v}\frac{\partial E_z}{\partial y} + \frac{\partial B_z}{\partial x}\right), \\
B_y &= \frac{1}{v}E_x + \frac{i\bar{\lambda}}{(n_c+\widehat{\wp}_z)}\left(\frac{1}{v}\frac{\partial E_z}{\partial x} - \frac{\partial B_z}{\partial y}\right).
\end{aligned}
\tag{218}
$$

These relations hold for any paraxial beam moving in the direction of $+z$-axis. These relationships can be written compactly as

$$\mathbb{F}(z) = \left(\frac{2ƛ}{(n_c + \widehat{\wp}_z)} (G_x \partial_x + G_y \partial_y) + G_z \right) \mathbb{F}(z), \tag{219}$$

where the matrices G_x, G_y and G_z are given as follows. Let

$$S_1 = \begin{pmatrix} 0 & 0 & 0 \\ 0 & 0 & -i \\ 0 & i & 0 \end{pmatrix}, \quad S_2 = \begin{pmatrix} 0 & 0 & i \\ 0 & 0 & 0 \\ -i & 0 & 0 \end{pmatrix}. \tag{220}$$

Then,

$$\begin{aligned} G_x &= \frac{1}{2}\begin{pmatrix} -S_2 & S_1 \\ -S_1 & -S_2 \end{pmatrix} = \frac{1}{2}\begin{pmatrix} 0 & 0 & -i & 0 & 0 & 0 \\ 0 & 0 & 0 & 0 & 0 & -i \\ i & 0 & 0 & 0 & i & 0 \\ 0 & 0 & 0 & 0 & 0 & -i \\ 0 & 0 & i & 0 & 0 & 0 \\ 0 & -i & 0 & i & 0 & 0 \end{pmatrix}, \\ G_y &= \frac{1}{2}\begin{pmatrix} S_1 & S_2 \\ -S_2 & S_1 \end{pmatrix} = \frac{1}{2}\begin{pmatrix} 0 & 0 & 0 & 0 & 0 & i \\ 0 & 0 & -i & 0 & 0 & 0 \\ 0 & i & 0 & -i & 0 & 0 \\ 0 & 0 & -i & 0 & 0 & 0 \\ 0 & 0 & 0 & 0 & 0 & -i \\ i & 0 & 0 & 0 & i & 0 \end{pmatrix}, \\ G_z &= \begin{pmatrix} 0 & 0 & 0 & 0 & 1 & 0 \\ 0 & 0 & 0 & -1 & 0 & 0 \\ 0 & 0 & 0 & 0 & 0 & 0 \\ 0 & -1 & 0 & 0 & 0 & 0 \\ 1 & 0 & 0 & 0 & 0 & 0 \\ 0 & 0 & 0 & 0 & 0 & 0 \end{pmatrix}, \end{aligned} \tag{221}$$

For these matrices we have followed the same notation as in (see Khan, 2016c, 2017c, 2017e; Mukunda et al., 1985b; Simon et al., 1987). The matrices G_x and G_y satisfy the following identities:

$$G_x G_y = G_y G_x, \quad G_x^2 + G_y^2 = 0, \quad G_x^3 = 0, \quad G_y^3 = 0. \tag{222}$$

The aforementioned relations in (218) are consistent with the results obtained using a group theoretical analysis of the Maxwell equations (Mukunda et al., 1983, 1985a, 1985b; Simon et al., 1986, 1987; Sudarshan et al., 1983).

Let us now consider a monochromatic paraxial beam to be incident from a homogeneous medium of constant refractive index n_c on an ideal linear optical system at the input xy-plane at z', pass through it, and emerge into that homogeneous medium at the output xy-plane at $z > z'$. The beam is moving through the entire system along the optic axis in the forward z-direction. Let the propagation of the beam through the optical system from the input plane to the output plane be described by the scalar wave optics as in (211). Since the output beam at z in the homogeneous medium should satisfy (219) we have

$$\begin{aligned} \mathbb{F}(z) &= \left(\frac{2\lambda\!\!\!^{-}}{(n_c + \widehat{\wp}_z)}(G_x \partial_x + G_y \partial_y) + G_z\right) \\ &\quad \times \widehat{\mathcal{G}}(\boldsymbol{r}_\perp, \widehat{\wp}_\perp; z, z')\mathbb{F}(z'). \end{aligned} \tag{223}$$

After some lengthy straightforward algebra (see (Khan, 2016c) for similar calculations), taking $\widehat{\wp}_z \approx n_c$, we arrive at the result

$$\begin{aligned} \mathbb{F}(z) &= \widehat{\mathcal{G}}\left(\boldsymbol{r}_\perp + \frac{\lambda\!\!\!^{-}}{n_c}\mathbf{G}_\perp, \widehat{\wp}_\perp; z, z'\right)\mathbb{F}(z') \\ &= \widehat{\mathcal{G}}(\mathbf{Q}_\perp, \widehat{\wp}_\perp; z, z')\mathbb{F}(z'). \end{aligned} \tag{224}$$

Thus, the z-evolution operator of the Maxwell vector wave optics is obtained from the z-evolution operator of the Helmholtz scalar wave optics by the substitution

$$\boldsymbol{r}_\perp \longrightarrow \mathbf{Q}_\perp = \boldsymbol{r}_\perp + \frac{\lambda\!\!\!^{-}}{n_c}\mathbf{G}_\perp. \tag{225}$$

This is the elegant Mukunda-Simon-Sudarshan matrix substitution rule for transition from the Helmholtz scalar wave optics to the Maxwell vector wave optics.

In summary, the MSS substitution rule for transition from the Helmholtz scalar wave optics to the Maxwell vector wave optics implies the following: Let the z-axis be the optic axis of a linear optical system. Let $\psi(\boldsymbol{r}_\perp, z')$ describe a light wave in the transverse plane at z' consistent with the Helmholtz scalar wave equation, and let

$$\psi(\boldsymbol{r}_\perp; z) = \widehat{\mathcal{G}}(\boldsymbol{r}_\perp, \widehat{\wp}_\perp; z, z')\psi(\boldsymbol{r}_\perp; z') \tag{226}$$

describe that light wave in the transverse plane at $z > z'$ after passing through the system in accordance with the Helmholtz equation. Then, the same light propagation through the system will be described in the Maxwell vector wave optics by

$$\mathbb{F}(z) = \widehat{\mathcal{G}}\left(\boldsymbol{r}_\perp + \frac{\lambda\!\!\!^{-}}{n_c}\mathbf{G}_\perp, \widehat{\wp}_\perp; z, z'\right)\mathbb{F}(z'), \tag{227}$$

for any $z > z'$. It is important to note that when the fields in $\mathbb{F}(z')$ satisfy the Maxwell equations the fields in $\mathbb{F}(z)$ will also satisfy the Maxwell equations (Simon et al., 1986).

6.2 A matrix differential operator form of the Mukunda-Simon-Sudarshan rule

Now, we shall derive a matrix differential operator form of the Mukunda-Simon-Sudarshan substitution rule for transition from the Helmholtz scalar wave optics to the Maxwell vector wave optics. Let us recall that the Taylor expansion of any function $f(x, y)$ about a point (a, b) is given by

$$\begin{aligned} f(x+a, y+b) &= f(x, y) + a\frac{\partial f(x, y)}{\partial x} + b\frac{\partial f(x, y)}{\partial y} \\ &+ \frac{1}{2!}\left(a^2\frac{\partial^2 f(x, y)}{\partial x^2} + 2ab\frac{\partial^2 f(x, y)}{\partial x \partial y} + b^2\frac{\partial^2 f(x, y)}{\partial y^2}\right) \\ &+ \cdots . \\ &= \left[1 + \left(a\frac{\partial}{\partial x} + b\frac{\partial}{\partial y}\right) + \frac{1}{2!}\left(a\frac{\partial}{\partial x} + b\frac{\partial}{\partial y}\right)^2 + \cdots\right] f(x, y) \\ &= \exp\left(a\frac{\partial}{\partial x} + b\frac{\partial}{\partial y}\right) f(x, y). \end{aligned} \tag{228}$$

We now apply the Taylor expansion to $\mathcal{G}(\widehat{\wp}_\perp, \boldsymbol{r}_\perp \to \mathbf{Q}_\perp; z, z')$ in (224). For $a = \lambda\!\!\!^{-} G_x/n_c$ and $b = \lambda\!\!\!^{-} G_y/n_c$, due to the identities in (222), the cubic and all higher power terms in the Taylor expansion are identically zero and we have

$$\begin{aligned} \mathcal{G}(\widehat{\wp}_\perp, \boldsymbol{r}_\perp \to \mathbf{Q}_\perp; z, z') &= \left\{\mathbf{1} + \frac{\lambda\!\!\!^{-}}{n_c}\mathbf{G}_\perp \cdot \nabla_\perp \right. \\ &\left. + \frac{1}{2!}\left(\frac{\lambda\!\!\!^{-}}{n_c}\mathbf{G}_\perp \cdot \nabla_\perp\right)^2\right\} \mathcal{G}(\widehat{\wp}_\perp, \boldsymbol{r}_\perp; z, z') \\ &= \exp\left(\frac{\lambda\!\!\!^{-}}{n_c}\mathbf{G}_\perp \cdot \nabla_\perp\right) \mathcal{G}(\widehat{\wp}_\perp, \boldsymbol{r}_\perp; z, z') \\ &= \widehat{\mathcal{D}}\, \mathcal{G}(\widehat{\wp}_\perp, \boldsymbol{r}_\perp; z, z'), \end{aligned} \tag{229}$$

where $\mathbf{1}$ is the 6×6 identity matrix. Thus, it is seen that the MSS substitution rule for transition from the Helmholtz scalar wave optics to the Maxwell vector wave optics is equivalent to the action of a matrix differential operator on the scalar z-evolution operator.

In the applications considered later we shall take the medium as vacuum with $n_c = 1$. With this in mind, we shall write down explicitly the matrix differential operator $\widehat{\mathcal{D}}$ in (229) taking $n_c = 1$:

$$
\begin{aligned}
\widehat{\mathcal{D}} &= \exp\left(\lambda\!\!\!^{-} \boldsymbol{G}_\perp \cdot \boldsymbol{\nabla}_\perp\right) = \begin{pmatrix} \widehat{\mathcal{D}}^{(11)} & \widehat{\mathcal{D}}^{(12)} \\ \widehat{\mathcal{D}}^{(21)} & \widehat{\mathcal{D}}^{(22)} \end{pmatrix}, \\
\widehat{\mathcal{D}}^{(11)} &= \begin{pmatrix} 1 + \frac{\lambda\!\!\!^{-2}}{8}\left(\frac{\partial^2}{\partial x^2} - \frac{\partial^2}{\partial y^2}\right) & \frac{\lambda\!\!\!^{-2}}{4}\frac{\partial^2}{\partial x \partial y} & -i\frac{\lambda\!\!\!^{-}}{2}\frac{\partial}{\partial x} \\ \frac{\lambda\!\!\!^{-2}}{4}\frac{\partial^2}{\partial x \partial y} & 1 - \frac{\lambda\!\!\!^{-2}}{8}\left(\frac{\partial^2}{\partial x^2} - \frac{\partial^2}{\partial y^2}\right) & -i\frac{\lambda\!\!\!^{-}}{2}\frac{\partial}{\partial y} \\ i\frac{\lambda\!\!\!^{-}}{2}\frac{\partial}{\partial x} & i\frac{\lambda\!\!\!^{-}}{2}\frac{\partial}{\partial y} & 1 \end{pmatrix}, \\
\widehat{\mathcal{D}}^{(12)} &= \begin{pmatrix} -\frac{\lambda\!\!\!^{-2}}{4}\frac{\partial^2}{\partial x \partial y} & \frac{\lambda\!\!\!^{-2}}{8}\left(\frac{\partial^2}{\partial x^2} - \frac{\partial^2}{\partial y^2}\right) & i\frac{\lambda\!\!\!^{-}}{2}\frac{\partial}{\partial y} \\ \frac{\lambda\!\!\!^{-2}}{8}\left(\frac{\partial^2}{\partial x^2} - \frac{\partial^2}{\partial y^2}\right) & \frac{\lambda\!\!\!^{-2}}{4}\frac{\partial^2}{\partial x \partial y} & -i\frac{\lambda\!\!\!^{-}}{2}\frac{\partial}{\partial x} \\ -i\frac{\lambda\!\!\!^{-}}{2}\frac{\partial}{\partial y} & i\frac{\lambda\!\!\!^{-}}{2}\frac{\partial}{\partial x} & 0 \end{pmatrix}, \\
\widehat{\mathcal{D}}^{(21)} &= \begin{pmatrix} \frac{\lambda\!\!\!^{-2}}{4}\frac{\partial^2}{\partial x \partial y} & -\frac{\lambda\!\!\!^{-2}}{8}\left(\frac{\partial^2}{\partial x^2} - \frac{\partial^2}{\partial y^2}\right) & -i\frac{\lambda\!\!\!^{-}}{2}\frac{\partial}{\partial y} \\ -\frac{\lambda\!\!\!^{-2}}{8}\left(\frac{\partial^2}{\partial x^2} - \frac{\partial^2}{\partial y^2}\right) & -\frac{\lambda\!\!\!^{-2}}{4}\frac{\partial^2}{\partial x \partial y} & i\frac{\lambda\!\!\!^{-}}{2}\frac{\partial}{\partial x} \\ i\frac{\lambda\!\!\!^{-}}{2}\frac{\partial}{\partial y} & -i\frac{\lambda\!\!\!^{-}}{2}\frac{\partial}{\partial x} & 0 \end{pmatrix}, \\
\widehat{\mathcal{D}}^{(22)} &= \begin{pmatrix} 1 + \frac{\lambda\!\!\!^{-2}}{8}\left(\frac{\partial^2}{\partial x^2} - \frac{\partial^2}{\partial y^2}\right) & \frac{\lambda\!\!\!^{-2}}{4}\frac{\partial^2}{\partial x \partial y} & -i\frac{\lambda\!\!\!^{-}}{2}\frac{\partial}{\partial x} \\ \frac{\lambda\!\!\!^{-2}}{4}\frac{\partial^2}{\partial x \partial y} & 1 - \frac{\lambda\!\!\!^{-2}}{8}\left(\frac{\partial^2}{\partial x^2} - \frac{\partial^2}{\partial y^2}\right) & -i\frac{\lambda\!\!\!^{-}}{2}\frac{\partial}{\partial y} \\ i\frac{\lambda\!\!\!^{-}}{2}\frac{\partial}{\partial x} & i\frac{\lambda\!\!\!^{-}}{2}\frac{\partial}{\partial y} & 1 \end{pmatrix}.
\end{aligned}
\tag{230}
$$

The matrix differential operator form of the MSS substitution rule is seen to be much easier to use than the direct substitution rule (Khan, 2023b).

6.3 Maxwell beams and cross polarization

Now, we shall study the Maxwell beams which are the vector generalizations of the scalar beams satisfying the Helmholtz equation following (Simon et al., 1987). Let a scalar plane wave $E_0 \exp(ikz)$ be incident at the transverse plane at $z = 0$ on the z-axis, or the optic axis, of a linear optical system with the amplitude transmittance function $\psi(\boldsymbol{r}_\perp; 0)$ where $\psi(\boldsymbol{r}_\perp; z)\exp(ikz)$ is a solution of the Helmholtz equation. The beam emerging from the system will have the transverse profile $E_0\psi(\boldsymbol{r}_\perp; 0)$ at $z = 0$. Then, after free propagation along the optic axis of the system, following the Helmholtz equation, the wave function at z will be $E_0\psi(\boldsymbol{r}_\perp; z)\exp(ikz)$. Hereafter, we shall drop the purely z-dependent overall phase factor $\exp(ikz)$ since it is independent of $\boldsymbol{r}_\perp$ and we are interested only in the transverse profile of the beam along the optic axis. Since the plane wave $E_0\exp(ikz)$ incident on the system at $z = 0$, with value E_0, is transformed into the wave function $E_0\psi(\boldsymbol{r}_\perp; z)$ in the transverse plane at z it is clear from (226) that, in this case,

$$\widehat{\mathcal{G}}(\boldsymbol{r}_\perp, \widehat{\wp}_\perp; z, 0) = \psi(\boldsymbol{r}_\perp; z). \tag{231}$$

We shall now look at the above system from the point of view of the Maxwell vector wave optics. Let us consider the plane wave propagating along the z-direction and passing through the system at $z = 0$ to be x-polarized, without loss of generality. Then its field at $z = 0$ can be represented by

$$\begin{pmatrix} \boldsymbol{E}(\boldsymbol{r}_\perp; 0) \\ c\boldsymbol{B}(\boldsymbol{r}_\perp; 0) \end{pmatrix} = \mathbb{F}(0) = E_0 \begin{pmatrix} 1 \\ 0 \\ 0 \\ 0 \\ 1 \\ 0 \end{pmatrix}. \tag{232}$$

Let us recall that for a plane electromagnetic wave in vacuum $c\boldsymbol{B} = \widehat{k} \times \boldsymbol{E}$ where $\widehat{k}$ is the unit vector in the direction of propagation. Then, it follows that the field of the Maxwell beam, corresponding to the scalar Helmholtz beam $\psi(\boldsymbol{r}_\perp; z)$, at z will be given by

$$\begin{aligned} \mathbb{F}(z) &= \widehat{\mathcal{G}}(\boldsymbol{r}_\perp + \lambda\!\!\!^{-}\mathbf{G}_\perp, \widehat{\wp}_\perp; z, 0)\mathbb{F}(0) \\ &= \psi(\boldsymbol{r}_\perp + \lambda\!\!\!^{-}\mathbf{G}_\perp; z)\mathbb{F}(0) \\ &= E_0\psi(x + \lambda\!\!\!^{-} G_x, y + \lambda\!\!\!^{-} G_y; z) \begin{pmatrix} 1 \\ 0 \\ 0 \\ 0 \\ 1 \\ 0 \end{pmatrix}, \end{aligned} \tag{233}$$

using the matrix substitution form of the MSS rule (233). Using the matrix differential operator form of the MSS rule we can write

$$\begin{aligned} \mathbb{F}(z) &= E_0 \left[\exp(\bar{\lambda} \mathbf{G}_\perp \cdot \nabla_\perp) \psi(\boldsymbol{r}_\perp; z) \right] \begin{pmatrix} 1 \\ 0 \\ 0 \\ 0 \\ 1 \\ 0 \end{pmatrix} \\ &= E_0 \left[\widehat{\mathcal{D}} \psi(\boldsymbol{r}_\perp; z) \right] \begin{pmatrix} 1 \\ 0 \\ 0 \\ 0 \\ 1 \\ 0 \end{pmatrix}. \end{aligned} \tag{234}$$

where $\widehat{\mathcal{D}}$ is defined in (230). At this point, one can use either the substitution form or the differential operator form of the MSS rule to evaluate $\mathbb{F}(z)$. In some cases the substitution rule is easy to apply, but the differential operator form is always straightforward and easy.

Let us use the differential operator form of the MSS rule to understand some general features of the electromagnetic field of a Maxwell beam. From (234) we have

$$\begin{aligned} E_x(\boldsymbol{r}_\perp; z) &= E_0 \left[\psi(\boldsymbol{r}_\perp; z) + \frac{\bar{\lambda}^2}{4} \left(\frac{\partial^2 \psi(\boldsymbol{r}_\perp; z)}{\partial x^2} - \frac{\partial^2 \psi(\boldsymbol{r}_\perp; z)}{\partial y^2} \right) \right], \\ E_y(\boldsymbol{r}_\perp; z) &= E_0 \left[\frac{\bar{\lambda}^2}{2} \left(\frac{\partial^2 \psi(\boldsymbol{r}_\perp; z)}{\partial x \partial y} \right) \right], \\ E_z(\boldsymbol{r}_\perp; z) &= E_0 \left[i\bar{\lambda} \left(\frac{\partial \psi(\boldsymbol{r}_\perp; z)}{\partial x} \right) \right], \\ cB_x(\boldsymbol{r}_\perp; z) &= E_0 \left[\frac{\bar{\lambda}^2}{2} \left(\frac{\partial^2 \psi(\boldsymbol{r}_\perp; z)}{\partial x \partial y} \right) \right], \\ cB_y(\boldsymbol{r}_\perp; z) &= E_0 \left[\psi(\boldsymbol{r}_\perp; z) - \frac{\bar{\lambda}^2}{4} \left(\frac{\partial^2 \psi(\boldsymbol{r}_\perp; z)}{\partial x^2} - \frac{\partial^2 \psi(\boldsymbol{r}_\perp; z)}{\partial y^2} \right) \right], \\ cB_z(\boldsymbol{r}_\perp; z) &= E_0 \left[i\bar{\lambda} \left(\frac{\partial \psi(\boldsymbol{r}_\perp; z)}{\partial y} \right) \right]. \end{aligned} \tag{235}$$

These are the formal expressions for the components of the field of the Maxwell beam characterized by a scalar wave function $\psi(\boldsymbol{r}_\perp; z)e^{ikz}$ satisfying the Helmholtz equation.

Note that E_x and B_y are the dominant components of the field of the Maxwell beam. In this sense the beam is x-polarized as a result of the initial plane wave incident on the transverse plane at $z = 0$ being x-polarized. Besides these dominant components the beam has smaller E_y and B_x known as the cross polarization components. The beam field has also the small longitudinal components E_z and B_z along the direction of propagation. These cross polarization and longitudinal components are seen to vanish as $\lambda \longrightarrow 0$.

We note the following interesting relations among the components:

$$E_y(\boldsymbol{r}_\perp; z) = cB_x(\boldsymbol{r}_\perp; z), \tag{236a}$$

$$\frac{\partial E_z(\boldsymbol{r}_\perp; z)}{\partial y} = c\frac{\partial B_z(\boldsymbol{r}_\perp; z)}{\partial x}, \tag{236b}$$

$$E_x(\boldsymbol{r}_\perp; z) + cB_y(\boldsymbol{r}_\perp; z) = 2E_0\psi(\boldsymbol{r}_\perp; z). \tag{236c}$$

It is to be noted that the equation (236a) relates the cross polarization components of the electric and magnetic fields. All the above three relations are found to be satisfied for the seven types of Maxwell beams considered later.

The presence of the longitudinal and cross polarization components is a general feature of all the Maxwell beams. Any laser beam has a finite extent in the directions transverse to the predominant direction of propagation. The finite transverse size of the beam requires the existence of all the three Cartesian components of the electric and magnetic fields due to the Maxwell equations the beam has to satisfy.

The total irradiance of any Maxwell beam at the transverse plane at z is given by

$$\begin{aligned}
\mathcal{I}(z) &= \int_{-\infty}^{\infty}\int_{-\infty}^{\infty} dxdy\ \boldsymbol{E}^*(\boldsymbol{r}_\perp; z)\cdot\boldsymbol{E}(\boldsymbol{r}_\perp; z) \\
&= \int_{-\infty}^{\infty}\int_{-\infty}^{\infty} dxdy\ |E_x(\boldsymbol{r}_\perp; z)|^2 + \int_{-\infty}^{\infty}\int_{-\infty}^{\infty} dxdy\ |E_y(\boldsymbol{r}_\perp; z)|^2 \\
&\quad + \int_{-\infty}^{\infty}\int_{-\infty}^{\infty} dxdy\ |E_z(\boldsymbol{r}_\perp; z)|^2 \\
&= \mathcal{I}_x(z) + \mathcal{I}_y(z) + \mathcal{I}_z(z).
\end{aligned} \tag{237}$$

7. Cross polarization in various Maxwell beams

In the previous Section, we briefly saw the derivation of the Mukunda-Simon-Sudarshan rule (MSS rule) for the transition from the Helmholtz scalar wave optics to the Maxwell vector wave optics. We showed that the action of the MSS rule can be implemented in two ways, namely, the matrix substitution form and the equivalent matrix differential operator form. Using the matrix differential operator form of the MSS rule we obtained the formal expressions for all the three components of the electric and magnetic fields of Maxwell beams and noted that the presence of the longitudinal and cross polarization components is a general feauture of any Maxwell beam. Now, we shall study the following seven types of light beams: Gaussian Beam, Anisotropic Gaussian Beam, Hermite-Gaussian Beam, Bessel Beam, Bessel-Gaussian Beam, Airy Beam, and the Anisotropic Airy Beam. For a lucid account of the functions involved in the description of all these beams see Lakshminarayanan & Varadharajan, 2015; (See also Abramowitz & Stegun, 2014; Arfken et al., 2012; Gradshteyn & Ryzhik, 2007; Spiegel & Liu, 1999).

Structured light fields have received enormous interest as they have unusual and interesting properties (Chávez-Cerda, 1999; Driben et al., 2013; Forbes, 2019, 2020; Forbes et al., 2021; Gómez-Correa et al., 2017; Hui-Chuan & Ji-Xiong, 2012; Jaimes-Nájera et al., 2020, 2022; Jia et al., 2023a, 2023b; Mandel & Wolf, 1995; Otte, 2020; Rogel-Salazar et al., 2014; Rubinsztein-Dunlop et al., 2017; Ugalde-Ontiveros et al., 2021a, 2021b).

7.1 Cross polarization in Gaussian Maxwell beams

The Gaussian Maxwell beam (Simon et al., 1987) is associated with the scalar Gaussian wave function

$$\psi_G(\boldsymbol{r}_\perp; z) = \sqrt{\frac{2}{\pi}}\, \frac{1}{\sigma(z)} \exp\left(\frac{ik}{2q(z)} r_\perp^2\right), \tag{238}$$

where the complex radius of curvature $q(z)$, the radius curvature of the phase front $R(z)$, and the beam width (or the spot size) $\sigma(z)$ are given by

$$\begin{aligned} \frac{1}{q(z)} &= \frac{1}{R(z)} + \frac{2i}{k\sigma^2(z)}, \\ \sigma(z) &= \sigma_0 \sqrt{1 + \left(\frac{2z}{k\sigma_0^2}\right)^2}, \\ R(z) &= z\left[1 + \left(\frac{k\sigma_0^2}{2z}\right)^2\right]. \end{aligned} \tag{239}$$

The function $\psi(\boldsymbol{r}_\perp, z)$ defined by (238)-(239) represents a scalar Gaussian beam which is a solution of the paraxial Helmhotz equation. Then, with

$$\mathbb{F}(0) = E_0 \begin{pmatrix} 1 \\ 0 \\ 0 \\ 0 \\ 1 \\ 0 \end{pmatrix}, \tag{240}$$

the Gaussin Maxwell beam is defined by

$$\mathbb{F}_G(z) = \sqrt{\frac{2}{\pi}} \frac{E_0}{\sigma(z)} \exp\left\{ \frac{ik}{2q(z)} \left[(x + \bar{\lambda} G_x)^2 + (y + \bar{\lambda} G_y)^2 \right] \right\} \begin{pmatrix} 1 \\ 0 \\ 0 \\ 0 \\ 1 \\ 0 \end{pmatrix}, \tag{241}$$

following the substitution form of the MSS rule (233).

Using the properties of the matrices G_x and G_y in (222), the exponential in (241) is readily obtained in a closed form. Thus, the expression in (241) simplifies to

$$\mathbb{F}_G(z) = \sqrt{\frac{2}{\pi}} \frac{E_0}{\sigma(z)} \exp\left(\frac{ik}{2q(z)} r_\perp^2 \right) \mathcal{M}_G \begin{pmatrix} 1 \\ 0 \\ 0 \\ 0 \\ 1 \\ 0 \end{pmatrix}, \tag{242}$$

where $\mathcal{M}_G$ is the matrix

$$\begin{aligned} \mathcal{M}_G &= 1_6 + \frac{i}{q} \boldsymbol{r}_\perp \cdot \mathbf{G}_\perp - \frac{1}{2q^2}(x^2 - y^2) G_x^2 - \frac{1}{q^2} xy G_x G_y \\ &= 1_6 + \begin{pmatrix} -\frac{x^2-y^2}{8q^2} & -\frac{xy}{4q^2} & \frac{x}{2q} & \frac{xy}{4q^2} & -\frac{x^2-y^2}{8q^2} & -\frac{y}{2q} \\ -\frac{xy}{4q^2} & \frac{x^2-y^2}{8q^2} & \frac{y}{2q} & -\frac{x^2-y^2}{8q^2} & -\frac{xy}{4q^2} & \frac{x}{2q} \\ -\frac{x}{2q} & -\frac{y}{2q} & 0 & \frac{y}{2q} & -\frac{x}{2q} & 0 \\ -\frac{xy}{4q^2} & \frac{x^2-y^2}{8q^2} & \frac{y}{2q} & -\frac{x^2-y^2}{8q^2} & -\frac{xy}{4q^2} & \frac{x}{2q} \\ \frac{x^2-y^2}{8q^2} & \frac{xy}{4q^2} & -\frac{x}{2q} & -\frac{xy}{4q^2} & \frac{x^2-y^2}{8q^2} & \frac{y}{2q} \\ -\frac{y}{2q} & \frac{x}{2q} & 0 & -\frac{x}{2q} & -\frac{y}{2q} & 0 \end{pmatrix}. \end{aligned} \tag{243}$$

Then, the field components are seen to be

$$\begin{aligned}
E_x(\boldsymbol{r}_\perp; z) &= E_0\left(1 - \frac{x^2 - y^2}{4q(z)^2}\right)\psi_G(\boldsymbol{r}_\perp; z),\\
E_y(\boldsymbol{r}_\perp; z) &= -E_0\left(\frac{xy}{2q(z)^2}\right)\psi_G(\boldsymbol{r}_\perp; z),\\
E_z(\boldsymbol{r}_\perp; z) &= -E_0\left(\frac{x}{2q(z)}\right)\psi_G(\boldsymbol{r}_\perp; z),\\
cB_x(\boldsymbol{r}_\perp; z) &= -E_0\left(\frac{xy}{2q(z)^2}\right)\psi_G(\boldsymbol{r}_\perp; z),\\
cB_y(\boldsymbol{r}_\perp; z) &= E_0\left(1 + \frac{x^2 - y^2}{4q(z)^2}\right)\psi_G(\boldsymbol{r}_\perp; z),\\
cB_z(\boldsymbol{r}_\perp; z) &= -E_0\left(\frac{y}{2q(z)}\right)\psi_G(\boldsymbol{r}_\perp; z).
\end{aligned} \tag{244}$$

The electric field $\boldsymbol{E}(\boldsymbol{r}_\perp; z)$ is seen to have a longitudinal component $E_z(\boldsymbol{r}_\perp; z)$ along the beam axis and a cross polarization component $E_y(\boldsymbol{r}_\perp; z)$, besides the dominant component $E_x(\boldsymbol{r}_\perp; z)$ in the direction of polarization. Similarly, the magnetic field $\boldsymbol{B}(\boldsymbol{r}_\perp; z)$ is seen to have a longitudinal z-component and a cross polarization x-component besides the dominant y-component. It is seen immediately that the cross polarization components $E_y(\boldsymbol{r}_\perp; z)$ and $B_x(\boldsymbol{r}_\perp; z)$ satisfy the relation in (236a) obtained from general considerations of any Maxwell beam. The other relations (236b)–(236c) are also easily verified. The cross polarization components vanish only along the two planes defined by $x = 0$ and $y = 0$, respectively. It is to be noted, that the intersection of these two planes is the z-axis, which is the optic axis.

Let us observe that

$$\psi_G^*(\boldsymbol{r}_\perp; z)\psi_G(\boldsymbol{r}_\perp; z) = \frac{2}{\pi}\frac{1}{\sigma^2(z)}\exp\left(-\frac{2}{\sigma^2(z)}r_\perp^2\right). \tag{245}$$

Thus, the evaluation of total irradiance (237) in the case of a Gaussian Maxwell beam involves the various moments of the Gaussian function in (245). It can be shown that

$$\begin{aligned}
\mathcal{I}_x(z) &= E_0^2\left[1 + \frac{1}{4}\left(\frac{\sigma^2(z)}{4\mid q(z)\mid^2}\right)^2\right]\\
\mathcal{I}_y(z) &= \frac{E_0^2}{4}\left(\frac{\sigma^2(z)}{4\mid q(z)\mid^2}\right)^2\\
\mathcal{I}_z(z) &= \frac{E_0^2\,\sigma^2(z)}{4\mid q(z)\mid^2}.
\end{aligned} \tag{246}$$

These results are consistent with the results of Simon et al. (1987).

7.2 Cross polarization in anisotropic Gaussian Maxwell beams

We shall now describe the cross polarization in anisotropic Gaussian Maxwell beams. These beams were brought to notice in 1983 as the exact solutions of the Helmholtz equation (Simon, 1983, 1985, 1987). They form the resonant modes of the Fabry-Perot laser resonators with astigmatic mirrors and are of immense interest (Seshadri, 2009). The anisotropic Gaussian Maxwell beam is associated with the scalar anisotropic Gaussian wave function

$$\psi_{aG}(\mathbf{r}_\perp; z) = \sqrt{\frac{2}{\pi}} \frac{1}{\sqrt{\sigma_1(z)\sigma_2(z)}} \exp\left(\frac{ik}{2q_1(z)}x^2 + \frac{ik}{2q_2(z)}y^2\right). \quad (247)$$

The complex radii of curvature, $q_1(z)$ and $q_2(z)$ and the radii of curvature of the phase fronts, $R_1(z)$ and $R_2(z)$ are, respectively, related as

$$\frac{1}{q_1(z)} = \frac{1}{R_1(z)} + \frac{2i}{k\sigma_1^2(z)}\frac{1}{q_2(z)} = \frac{1}{R_2(z)} + \frac{2i}{k\sigma_2^2(z)}, \quad (248)$$

where $\sigma_1(z)$ and $\sigma_2(z)$ are the unequal beam widths along two mutually perpendicular directions orthogonal to the beam axis (z-axis). As in the case of the Gaussian beams, the field of an anisotropic Gaussian Maxwell beam can be evaluated exactly using the MSS matrix substitution rule. For other beams, it will be easier to use the differential operator form of the MSS matrix substitution rule.

Then, with

$$\mathbb{F}(0) = E_0 \begin{pmatrix} 1 \\ 0 \\ 0 \\ 0 \\ 1 \\ 0 \end{pmatrix}, \quad (249)$$

the anisotropic Gaussin Maxwell beam is defined by

$$\begin{aligned} \mathbb{F}_{aG}(z) &= \sqrt{\frac{2}{\pi}} \frac{E_0}{\sqrt{\sigma_1(z)\sigma_2(z)}} \\ &\quad \times \exp\left[\frac{ik}{2q_1(z)}(x + \lambda\!\!\!^{-} G_x)^2 + \frac{ik}{2q_2(z)}(y + \lambda\!\!\!^{-} G_y)^2\right] \begin{pmatrix} 1 \\ 0 \\ 0 \\ 0 \\ 1 \\ 0 \end{pmatrix} \quad (250) \\ &= \sqrt{\frac{2}{\pi}} \frac{E_0}{\sqrt{\sigma_1(z)\sigma_2(z)}} \exp\left(\frac{ik}{2q_1(z)}x^2 + \frac{ik}{2q_2(z)}y^2\right) \mathcal{M}_{aG} \begin{pmatrix} 1 \\ 0 \\ 0 \\ 0 \\ 1 \\ 0 \end{pmatrix}, \end{aligned}$$

following the substitution form of the MSS rule (233). The matrix $\mathcal{M}_{aG}$ is

$$
\begin{aligned}
\mathcal{M}_{aG} &= 1_6 + \frac{i}{q_1} x G_x + \frac{i}{q_2} y G_y - \frac{1}{2q_1^2} x^2 G_x^2 - \frac{1}{2q_2^2} y^2 G_y^2 - \frac{1}{q_1 q_2} xy G_x G_y \\
&= \begin{pmatrix} \mathcal{M}_{aG}^{(11)} & \mathcal{M}_{aG}^{(12)} \\ \mathcal{M}_{aG}^{(21)} & \mathcal{M}_{aG}^{(22)} \end{pmatrix},
\end{aligned}
\tag{251}
$$

with

$$
\begin{aligned}
\mathcal{M}_{aG}^{(11)} &= \begin{pmatrix} 1 + c_1 - \frac{1}{8}\left(\frac{x^2}{q_1^2} - \frac{y^2}{q_2^2}\right) & -\frac{xy}{4q_1 q_2} & \frac{x}{2q_1} \\ -\frac{xy}{4q_1 q_2} & 1 - c_1 + \frac{1}{8}\left(\frac{x^2}{q_1^2} - \frac{y^2}{q_2^2}\right) & \frac{y}{2q_2} \\ -\frac{x}{2q_1} & -\frac{y}{2q_2} & 1 \end{pmatrix}, \\
\mathcal{M}_{aG}^{(12)} &= \begin{pmatrix} \frac{xy}{4q_1 q_2} & c_1 - \frac{1}{8}\left(\frac{x^2}{q_1^2} - \frac{y^2}{q_2^2}\right) & -\frac{y}{2q_2} \\ c_1 - \frac{1}{8}\left(\frac{x^2}{q_1^2} - \frac{y^2}{q_2^2}\right) & -\frac{xy}{4q_1 q_2} & \frac{x}{2q_1} \\ \frac{y}{2q_2} & -\frac{x}{2q_1} & 0 \end{pmatrix}, \\
\mathcal{M}_{aG}^{(21)} &= \begin{pmatrix} -\frac{xy}{4q_1 q_2} & -c_1 + \frac{1}{8}\left(\frac{x^2}{q_1^2} - \frac{y^2}{q_2^2}\right) & \frac{y}{2q_2} \\ -c_1 + \frac{1}{8}\left(\frac{x^2}{q_1^2} - \frac{y^2}{q_2^2}\right) & \frac{xy}{4q_1 q_2} & -\frac{x}{2q_1} \\ -\frac{y}{2q_2} & \frac{x}{2q_1} & 0 \end{pmatrix}, \\
\mathcal{M}_{aG}^{(22)} &= \begin{pmatrix} 1 + c_1 - \frac{1}{8}\left(\frac{x^2}{q_1^2} - \frac{y^2}{q_2^2}\right) & -\frac{xy}{4q_1 q_2} & \frac{x}{2q_1} \\ -\frac{xy}{4q_1 q_2} & 1 - c_1 + \frac{1}{8}\left(\frac{x^2}{q_1^2} + \frac{y^2}{q_2^2}\right) & \frac{y}{2q_2} \\ -\frac{x}{2q_1} & -\frac{y}{2q_2} & 1 \end{pmatrix},
\end{aligned}
\tag{252}
$$

where $c_1 = i\frac{\bar{\lambda}}{8}\left(\frac{1}{q_1} - \frac{1}{q_2}\right)$. The field components are

$$\begin{aligned}
E_x(\boldsymbol{r}_\perp; z) &= E_0\left[(1 + 2c_1) - \frac{1}{4}\left(\frac{x^2}{q_1^2} - \frac{y^2}{q_2^2}\right)\right]\psi_{aG}(\boldsymbol{r}_\perp; z),\\
E_y(\boldsymbol{r}_\perp; z) &= -E_0\left(\frac{xy}{2q_1q_2}\right)\psi_{aG}(\boldsymbol{r}_\perp; z),\\
E_z(\boldsymbol{r}_\perp; z) &= -E_0\left(\frac{x}{q_1}\right)\psi_{aG}(\boldsymbol{r}_\perp; z),\\
cB_x(\boldsymbol{r}_\perp; z) &= -E_0\left(\frac{xy}{2q_1q_2}\right)\psi_{aG}(\boldsymbol{r}_\perp; z),\\
cB_y(\boldsymbol{r}_\perp; z) &= E_0\left[(1 - 2c_1) + \frac{1}{4}\left(\frac{x^2}{q_1^2} - \frac{y^2}{q_2^2}\right)\right]\psi_{aG}(\boldsymbol{r}_\perp; z),\\
cB_z(\boldsymbol{r}_\perp; z) &= -E_0\left(\frac{y}{q_2}\right)\psi_{aG}(\boldsymbol{r}_\perp; z).
\end{aligned} \tag{253}$$

We note that the electric and magnetic fields of an anisotropic Gaussian Maxwell beam have longitudinal and cross polarization components and satisfy the relations in (236a)–(236c) obtained from general considerations of the paraxial Maxwell beams. The results for the Gaussian beams (Mukunda et al., 1983, 1985a; Simon et al., 1986, 1987) are completely recovered from our results for the anisotropic Gaussian beams by equating $q_1(z) = q_2(z) = q(z)$ and $\sigma_1(z) = \sigma_2(z) = \sigma(z)$.

To evaluate the irradiance integrals in (237), we note that $\psi^*_{aG}(\boldsymbol{r}_\perp; z)\psi_{aG}(\boldsymbol{r}_\perp; z)$ is a Gaussian function with unequal widths $\sigma_1(z)$ and $\sigma_2(z)$:

$$\begin{aligned}
&\psi^*_{aG}(\boldsymbol{r}_\perp; z)\psi_{aG}(\boldsymbol{r}_\perp; z)\\
&= \frac{2}{\pi}\frac{1}{\sigma_1(z)\sigma_2(z)}\exp\left[-2\left(\frac{x^2}{\sigma_1^2(z)} + \frac{y^2}{\sigma_2^2(z)}\right)\right].
\end{aligned} \tag{254}$$

The integrals in (237) involve the various moments of the Gaussian function in (254). It can be shown that

$$
\begin{aligned}
\mathcal{I}_x(z) &= E_0^2 \Biggl\{ 1 - \left(\frac{1}{\sigma_1^2(z)} - \frac{1}{\sigma_2^2(z)} \right) + 4c_1^* c_1 \\
&\quad - \frac{1}{4}\{1 + (c_1 + c_1^*)\} \left(\frac{\sigma_1^2(z)}{|\, q_1(z) \,|^2} - \frac{\sigma_2^2(z)}{|\, q_2(z) \,|^2} \right) \\
&\quad + \frac{1}{256}\Biggl[3\frac{\sigma_1^4(z)}{|\, q_1(z) \,|^4} + 3\frac{\sigma_2^4(z)}{|\, q_2(z) \,|^4} \\
&\quad - \left(\frac{1}{q_1^{*2}(z)\, q_2^2(z)} + \frac{1}{q_1^2(z)\, q_2^{*2}(z)} \right) \sigma_1^2(z) \sigma_2^2(z) \Biggr] \Biggr\}, \\
\mathcal{I}_y(z) &= \frac{E_0^2}{4} \left[\frac{\sigma_1(z)\, \sigma_2(z)}{4 \,|\, q_1(z) \,|\,|\, q_2(z) \,|} \right]^2, \\
\mathcal{I}_z(z) &= \frac{\sigma_1^2(z)}{4 \,|\, q_1(z) \,|^2}.
\end{aligned}
\tag{255}
$$

It is seen that in the limit $\sigma_1(z) = \sigma_2(z) = \sigma(z)$ and $q_1(z) = q_2(z) = q(z)$ we have $c_1 = 0$ and the results for the anisotropic Gaussian Maxwell beam become identical to the results for the Gaussian Maxwell beam. Thus our results for the anisotropic Gaussian beams (Khan, 2024) are generalizations of the results for the isotropic Gaussian beams (Simon et al., 1987).

7.3 Cross polarization in Hermite-Gaussian Maxwell beams

Apart from the Gaussian beam solutions, the Helmholtz equation permits other classes of solutions. One such class of solutions is the Hermite-Gaussian beams (Erikson & Singh, 1994). It is to be noted that the laser resonators bounded by spherical mirrors have Gaussian beams as their fundamental modes and Hermite-Gaussian beams as higher-order modes. Hermite-Gaussian beams are a family of structurally stable laser modes, which have rectangular symmetry along the propagation axis. An interesting property of these beams is that they propagate in a shape-invariant manner, not only under free propagation but also through any first-order systems (Simon & Mukunda, 1998). Such beams play a fundamental role in the first-order optics, enabling the use of certain group theoretic approaches (Simon & Mukunda, 1998). Hence, the Hermite-Gaussian beams have drawn immense interest (Saghafi et al., 2001).

The scalar Hermite-Gaussian beam is given by

$$
\begin{aligned}
\psi_{HG}(\boldsymbol{r}_\perp; z) &= H_m\left(\frac{\sqrt{2}x}{\sigma(z)}\right) H_n\left(\frac{\sqrt{2}y}{\sigma(z)}\right) \psi_G(\boldsymbol{r}_\perp; z) \\
&= \sqrt{\frac{2}{\pi}} \frac{1}{\sigma(z)} H_m\left(\frac{\sqrt{2}x}{\sigma(z)}\right) H_n\left(\frac{\sqrt{2}y}{\sigma(z)}\right) \exp\left(\frac{ik}{2q(z)} r_\perp^2\right),
\end{aligned}
\tag{256}
$$

where $H_m(x)$ and $H_n(y)$ are the Hermite polynomials of degree m and n respectively and ψ_G is the scalar Gaussian wave function in (238). The beam spot size in the plane that intersects the beam axis at z is given by $\sigma(z)$. In writing the expression for the Hermite-Gaussian beam we have left out the purely z-dependent phase factors independent of (x, y) (For more details, see: Conry et al., 2012; Erikson & Singh, 1994; Saghafi et al., 2001; Simon & Mukunda, 1998).

Then, with

$$\mathbb{F}(0) = E_0 \begin{pmatrix} 1 \\ 0 \\ 0 \\ 0 \\ 1 \\ 0 \end{pmatrix}, \tag{257}$$

the Hermite-Gaussian Maxwell beam is defined by

$$\begin{aligned} \mathbb{F}_{HG}(z) &= \sqrt{\frac{2}{\pi}} \frac{E_0}{\sigma(z)} H_m\left(\frac{\sqrt{2}}{\sigma(z)}(x + \lambda\!\!\!^{-} G_x)\right) H_n\left(\frac{\sqrt{2}}{\sigma(z)}(y + \lambda\!\!\!^{-} G_y)\right) \\ &\quad \times \exp\left\{\frac{ik}{2q(z)}\left[(x + \lambda\!\!\!^{-} G_x)^2 + (y + \lambda\!\!\!^{-} G_y)^2\right]\right\} \begin{pmatrix} 1 \\ 0 \\ 0 \\ 0 \\ 1 \\ 0 \end{pmatrix}. \end{aligned} \tag{258}$$

following the substitution form of the MSS rule (233).

It is possible to obtain the closed form expressions for the electromagnetic field components of the Hermite-Gaussian Maxwell beam defined by $\mathbb{F}(z)$ in (258). This is done by using the standard properties of the exponential and the Hermite functions along with the identities in (222). After considerable but straightforward algebra, we find that

$$\mathbb{F}_{HG}(z) = \sqrt{\frac{2}{\pi}} \frac{E_0}{\sigma(z)} \exp\left(\frac{ik}{2q(z)} r_\perp^2\right) \mathcal{M}_{HG} \begin{pmatrix} 1 \\ 0 \\ 0 \\ 0 \\ 1 \\ 0 \end{pmatrix}, \tag{259}$$

where $\mathcal{M}_{HG}$ is a matrix written as the sum of two matrices

$$\mathcal{M}_{HG} = H_m H_n \mathcal{M}_G + \mathcal{M}_H, \tag{260}$$

with subscripts 'G' and 'H' indicating the 'Gaussian' and 'Hermite' contributions, respectively. The matrix $\mathcal{M}_G$ is the same as in (243) obtained for the Gaussian Maxwell beam. The matrix $\mathcal{M}_H$ is given by

$$\begin{aligned}
\mathcal{M}_H &= \begin{pmatrix} \mathcal{M}_H^{(11)} & \mathcal{M}_H^{(12)} \\ \mathcal{M}_H^{(21)} & \mathcal{M}_H^{(22)} \end{pmatrix}, \\
\mathcal{M}_H^{(11)} &= \begin{pmatrix} A & B & -\frac{i}{2}\lambda\!\!\!{}^{-} H'_m H_n \\ B & -A & -\frac{i}{2}\lambda\!\!\!{}^{-} H_m H'_n \\ -\frac{i}{2}\lambda\!\!\!{}^{-} H'_m H_n & -\frac{i}{2}\lambda\!\!\!{}^{-} H_m H'_n & 0 \end{pmatrix}, \\
\mathcal{M}_H^{(12)} &= \begin{pmatrix} A & B & -\frac{i}{2}\lambda\!\!\!{}^{-} H'_m H_n \\ B & -A & -\frac{i}{2}\lambda\!\!\!{}^{-} H_m H'_n \\ -\frac{i}{2}\lambda\!\!\!{}^{-} H'_m H_n & -\frac{i}{2}\lambda\!\!\!{}^{-} H_m H'_n & 0 \end{pmatrix}, \\
\mathcal{M}_H^{(21)} &= \begin{pmatrix} B & -A & -\frac{i}{2}\lambda\!\!\!{}^{-} H_m H'_n \\ -A & -B & \frac{i}{2}\lambda\!\!\!{}^{-} H'_m H_n \\ \frac{i}{2}\lambda\!\!\!{}^{-} H_m H'_n & -\frac{i}{2}\lambda\!\!\!{}^{-} H'_m H_n & 0 \end{pmatrix}, \\
\mathcal{M}_H^{(22)} &= \begin{pmatrix} -A & B & -\frac{i}{2}\lambda\!\!\!{}^{-} H'_m H_n \\ B & A & -\frac{i}{2}\lambda\!\!\!{}^{-} H_m H'_n \\ \frac{i}{2}\lambda\!\!\!{}^{-} H'_m H_n & \frac{i}{2}\lambda\!\!\!{}^{-} H_m H'_n & 0 \end{pmatrix},
\end{aligned} \tag{261}$$

where

$$\begin{aligned}
A &= \frac{1}{8}\lambda\!\!\!{}^{-2}\left(H''_m H_n - H_m H''_n\right) - \frac{i}{4q}\left(x H'_m H_n - y H_m H'_n\right), \\
B &= \frac{1}{4}\lambda\!\!\!{}^{-2} H'_m H'_n - \frac{i}{4q}\left(x H_m H'_n + y H'_m H_n\right), \\
H'_m &= \frac{d}{dx} H_m\left(\frac{\sqrt{2}}{\sigma(z)}x\right) = 2m\frac{\sqrt{2}}{\sigma(z)} H_{m-1}\left(\frac{\sqrt{2}}{\sigma(z)}x\right) \\
H'_n &= \frac{d}{dy} H_n\left(\frac{\sqrt{2}}{\sigma(z)}y\right) = 2n\frac{\sqrt{2}}{\sigma(z)} H_{n-1}\left(\frac{\sqrt{2}}{\sigma(z)}y\right).
\end{aligned} \tag{262}$$

The electromagnetic field components are

$$
\begin{aligned}
E_x(\boldsymbol{r}_\perp;\, z) &= E_0 \psi_G(\boldsymbol{r}_\perp;\, z)\Bigg[H_m H_n\left(1 - \frac{x^2-y^2}{4q^2}\right) \\
&\quad + \left(\frac{1}{4}\lambda\!\!\!^{-2}\left(H_m'' H_n - H_m H_n''\right) - \frac{i}{2q}\left(x H_m' H_n - y H_m H_n'\right)\right)\Bigg], \\
E_y(\boldsymbol{r}_\perp;\, z) &= E_0 \psi_G(\boldsymbol{r}_\perp;\, z)\Bigg[H_m H_n\left(-\frac{xy}{2q^2}\right) \\
&\quad + \left(\frac{1}{2}\lambda\!\!\!^{-2} H_m' H_n' - \frac{i}{2q}\left(x H_m H_n' + y H_m' H_n\right)\right)\Bigg], \\
E_z(\boldsymbol{r}_\perp;\, z) &= E_0 \psi_G(\boldsymbol{r}_\perp;\, z)\Bigg[H_m H_n\left(-\frac{x}{q}\right) + \left(i\lambda\!\!\!^{-} H_m' H_n\right)\Bigg] \\
cB_x(\boldsymbol{r}_\perp;\, z) &= E_0 \psi_G(\boldsymbol{r}_\perp;\, z)\Bigg[H_m H_n\left(-\frac{xy}{2q^2}\right) \\
&\quad + \left(\frac{1}{2}\lambda\!\!\!^{-2} H_m' H_n' - \frac{i}{2q}\left(x H_m H_n' + y H_m' H_n\right)\right)\Bigg], \\
cB_y(\boldsymbol{r}_\perp;\, z) &= E_0 \psi_G(\boldsymbol{r}_\perp;\, z)\Bigg[H_m H_n\left(1 + \frac{x^2-y^2}{4q^2}\right) \\
&\quad - \left(\frac{1}{4}\lambda\!\!\!^{-2}\left(H_m'' H_n - H_m H_n''\right) - \frac{i}{2q}\left(x H_m' H_n - y H_m H_n'\right)\right)\Bigg], \\
cB_z(\boldsymbol{r}_\perp;\, z) &= E_0 \psi_G(\boldsymbol{r}_\perp;\, z)\Bigg[H_m H_n\left(-\frac{y}{q}\right) + \lambda\!\!\!^{-} H_m H_n'\Bigg].
\end{aligned}
\tag{263}
$$

The results for the electric field components are consistent with those of Conry et al. (2012). We have derived the magnetic field components also. The Gaussian beam is a special case of the Hermite-Gaussian beam.

7.4 Cross polarization in Bessel Maxwell beams

It is known that the Helmholtz equation permits exact solutions that can propagate as beams without broadening and without divergence. These solutions are the Bessel functions. The solution of the lowest order have the transverse field distribution given by the $J_0\,(r_\perp;\, z)$, the zeroth order Bessel function of $r_\perp = |\boldsymbol{r}_\perp| = \sqrt{x^2 + y^2}$. But $J_0\,(r_\perp;\, z)$ is not square integrable, which implies that it would reqquire an infinite amount of energy to realize such beams! However, the Bessel beams have an enormous academic interest (Gómez-Correa et al., 2017; Wolf, 1988).

To obtain a Bessel Maxwell beam let us consider the simplest form of the scalar Bessel beam (Mishra, 1991; Sheppard et al., 2009):

$$
\psi_B\,(\boldsymbol{r}_\perp;\, z) = J_0\,(\alpha r_\perp), \tag{264}
$$

where $J_0(\alpha r_\perp)$ is the zeroth order Bessel function. It is possible to use the MSS matrix substitution rule (233), along with the various properties of the

Bessel function, to derive the Bessel Maxwell beam. However, it is easier to use the differential operator form of the MSS rule (234). Using (234) leads to the definition of the Bessel Maxwell beam as

$$\begin{aligned} \mathbb{F}_B(z) &= E_0 \mathcal{M}_B \begin{pmatrix} 1 \\ 0 \\ 0 \\ 0 \\ 1 \\ 0 \end{pmatrix} \\ \mathcal{M}_B &= \left(\mathcal{M}_{B0} J_0\left(\alpha r_\perp\right) + \mathcal{M}_{B1} J_1\left(\alpha r_\perp\right) \right), \end{aligned} \tag{265}$$

where

$$\mathcal{M}_{B0} = 1_6 - \frac{\alpha^2 \bar{\lambda}^2}{4 r_\perp^2} \mathcal{M}_{xy}, \tag{266a}$$

$$\mathcal{M}_{xy} = \begin{pmatrix} x^2 - y^2 & 2xy & 0 & -2xy & x^2 - y^2 & 0 \\ 2xy & -(x^2 - y^2) & 0 & x^2 - y^2 & 2xy & 0 \\ 0 & 0 & 0 & 0 & 0 & 0 \\ 2xy & -(x^2 - y^2) & 0 & x^2 - y^2 & 2xy & 0 \\ -(x^2 - y^2) & -2xy & 0 & 2xy & -(x^2 - y^2) & 0 \\ 0 & 0 & 0 & 0 & 0 & 0 \end{pmatrix}, \tag{266b}$$

and

$$\mathcal{M}_{B1} = \frac{\alpha \bar{\lambda}^2}{2} \begin{pmatrix} \frac{x^2 - y^2}{r_\perp^3} & \frac{2xy}{r_\perp^3} & \frac{i}{\bar{\lambda}} \frac{x}{r_\perp} & -\frac{2xy}{r_\perp^3} & \frac{x^2 - y^2}{r_\perp^3} & -\frac{i}{\bar{\lambda}} \frac{y}{r_\perp} \\ \frac{2xy}{r_\perp^3} & -\frac{x^2 - y^2}{r_\perp^3} & \frac{i}{\bar{\lambda}} \frac{y}{r_\perp} & \frac{x^2 - y^2}{r_\perp^3} & \frac{2xy}{r_\perp^3} & \frac{i}{\bar{\lambda}} \frac{x}{r_\perp} \\ -\frac{i}{\bar{\lambda}} \frac{x}{r_\perp} & -\frac{i}{\bar{\lambda}} \frac{y}{r_\perp} & 0 & \frac{i}{\bar{\lambda}} \frac{y}{r_\perp} & -\frac{i}{\bar{\lambda}} \frac{x}{r_\perp} & 0 \\ \frac{2xy}{r_\perp^3} & -\frac{x^2 - y^2}{r_\perp^3} & \frac{i}{\bar{\lambda}} \frac{y}{r_\perp} & \frac{x^2 - y^2}{r_\perp^3} & \frac{2xy}{r_\perp^3} & \frac{i}{\bar{\lambda}} \frac{x}{r_\perp} \\ -\frac{x^2 - y^2}{r_\perp^3} & -\frac{2xy}{r_\perp^3} & -\frac{i}{\bar{\lambda}} \frac{x}{r_\perp} & \frac{2xy}{r_\perp^3} & -\frac{x^2 - y^2}{r_\perp^3} & \frac{i}{\bar{\lambda}} \frac{y}{r_\perp} \\ -\frac{i}{\bar{\lambda}} \frac{y}{r_\perp} & \frac{i}{\bar{\lambda}} \frac{x}{r_\perp} & 0 & -\frac{i}{\bar{\lambda}} \frac{x}{r_\perp} & -\frac{i}{\bar{\lambda}} \frac{y}{r_\perp} & 0 \end{pmatrix}. \tag{267}$$

The components of the field of the Bessel Maxwell beam are as follows:

$$\begin{aligned} E_x(\boldsymbol{r}_\perp; z) &= E_0\left[\left(1 - \frac{\alpha^2 \bar{\lambda}^2}{2}\frac{x^2 - y^2}{r_\perp^2}\right)J_0(\alpha r_\perp) + \alpha \bar{\lambda}^2 \frac{x^2 - y^2}{r_\perp^3}J_1(\alpha r_\perp)\right], \\ E_y(\boldsymbol{r}_\perp; z) &= E_0\left(-\alpha^2 \bar{\lambda}^2 \frac{xy}{r_\perp^2}J_0(\alpha r_\perp) + \alpha \bar{\lambda}^2 \frac{xy}{r_\perp^3}J_1(\alpha r_\perp)\right), \\ E_z(\boldsymbol{r}_\perp; z) &= E_0\left(-i\alpha \bar{\lambda}\frac{x}{r_\perp}J_1(\alpha r_\perp)\right), \\ cB_x(\boldsymbol{r}_\perp; z) &= E_0\left(-\alpha^2 \bar{\lambda}^2 \frac{xy}{r_\perp^2}J_0(\alpha r_\perp) + \alpha \bar{\lambda}^2 \frac{xy}{r_\perp^3}J_1(\alpha r_\perp)\right), \\ cB_y(\boldsymbol{r}_\perp; z) &= E_0\left[\left(1 + \frac{\alpha^2 \bar{\lambda}^2}{2}\frac{x^2 - y^2}{r_\perp^2}\right)J_0(\alpha r_\perp) - \alpha \bar{\lambda}^2 \frac{x^2 - y^2}{r_\perp^3}J_1(\alpha r_\perp)\right], \\ cB_z(\boldsymbol{r}_\perp; z) &= E_0\left(-i\alpha \bar{\lambda}\frac{y}{r_\perp}J_1(\alpha r_\perp)\right). \end{aligned} \tag{268}$$

The fields thus obtained are consistent with the ones derived using the direct solutions in (Mishra, 1991; Sheppard et al., 2009; Wolf, 1988).

7.5 Cross polarization in Bessel-Gaussian Maxwell beams

Mathematically, the simplest Bessel beams are the ones, whose transverse field distribution is the zeroth order Bessel function, which is not square integrable. Such beams are not physical, as they would require an infinite amount of energy to produce them. These nonphysical beams have drawn substantial academic interest (Gori et al., 1987; Li et al., 2004; Mishra, 1991; Sheppard et al., 2009; Wolf, 1988). There have been attempts to create modified Bessel beams. Such beams carry finite power and are almost nondiffracting, as they can propagate over a long range without significant divergence. One such type is the Bessel-Gaussian beam (also called as Bessel-Gauss beam). The field amplitude distribution of the simplest scalar Bessel-Gaussian beam can be expressed in the form (Li et al., 2004)

$$\begin{aligned} \psi_{BG}(\boldsymbol{r}_\perp; z) &= J_0(\alpha r_\perp)\psi_G(\boldsymbol{r}_\perp; z) \\ &= \sqrt{\frac{2}{\pi}}\frac{1}{\sigma(z)}J_0(\alpha r_\perp)\exp\left(\frac{ik}{2q(z)}r_\perp^2\right). \end{aligned} \tag{269}$$

Then, with

$$\mathbb{F}(0) = E_0 \begin{pmatrix} 1 \\ 0 \\ 0 \\ 0 \\ 1 \\ 0 \end{pmatrix}, \tag{270}$$

using the differential operator form of the MSS rule (234) leads to the Bessel-Gaussian Maxwell beam defined by

$$
\begin{aligned}
\mathbb{F}_{BG}(z) &= \sqrt{\frac{2}{\pi}}\frac{1}{\sigma(z)}\exp\left(\frac{ik}{2q(z)}r_\perp^2\right)\mathcal{M}_{BG}\begin{pmatrix}1\\0\\0\\0\\1\\0\end{pmatrix}, \\
\mathcal{M}_{BG} &= (\mathcal{M}_G + \mathcal{M}_{B0} - 1_6)J_0(\alpha r_\perp) \\
&\quad + (\mathcal{M}_{B1} + \mathcal{M}_{BG1})J_1(\alpha r_\perp),
\end{aligned}
\tag{271}
$$

where the matrices $\mathcal{M}_G$, $\mathcal{M}_{B0}$, and $\mathcal{M}_{B1}$ are the same as in (243), (266), and (267), respectively, and

$$
\mathcal{M}_{BG1} = \frac{ikr}{\alpha q}\mathcal{M}_{xy}, \tag{272}
$$

with the matrix $\mathcal{M}_{xy}$ given in (266b).

After considerable but straightforward algebra, the expressions for the the field components are obtained as

$$
\begin{aligned}
E_x(\boldsymbol{r}_\perp; z) &= E_0\psi_G(\boldsymbol{r}_\perp; z)\left[\left(1 - \frac{x^2-y^2}{4q^2} - \frac{\alpha^2\bar{\lambda}^2}{2}\frac{x^2-y^2}{r_\perp^2}\right)J_0(\alpha r_\perp)\right. \\
&\quad \left. + \left(\alpha\bar{\lambda}^2\frac{x^2-y^2}{r_\perp^3} - i\frac{\alpha\bar{\lambda}^2 k}{2q}\frac{x^2-y^2}{r_\perp} - i\frac{\alpha k\bar{\lambda}^2}{q}\frac{xy}{r_\perp}\right)J_1(\alpha r_\perp)\right] \\
E_y(\boldsymbol{r}_\perp; z) &= E_0\psi_G(\boldsymbol{r}_\perp; z)\left(-\frac{xy}{2q^2} - \alpha^2\bar{\lambda}^2\frac{xy}{r_\perp^2}\right)J_0(\alpha r_\perp) \\
&\quad + \left(\alpha\bar{\lambda}^2\frac{xy}{r_\perp^3} - i\frac{\alpha k\bar{\lambda}^2}{q}\frac{xy}{r_\perp}\right)J_1(\alpha r_\perp) \\
E_z(\boldsymbol{r}_\perp; z) &= E_0\psi_G(\boldsymbol{r}_\perp; z)\left(-\frac{x}{q}J_0(\alpha r_\perp) - i\alpha\bar{\lambda}\frac{x}{r_\perp}J_1(\alpha r_\perp)\right), \\
cB_x(\boldsymbol{r}_\perp; z) &= E_0\psi_G(\boldsymbol{r}_\perp; z)\left(-\frac{xy}{2q^2} - \alpha^2\bar{\lambda}^2\frac{xy}{r_\perp^2}\right)J_0(\alpha r_\perp) \\
&\quad + \left(\alpha\bar{\lambda}^2\frac{xy}{r_\perp^3} - i\frac{\alpha k\bar{\lambda}^2}{q}\frac{xy}{r_\perp}\right)J_1(\alpha \boldsymbol{r}) \\
cB_y(\boldsymbol{r}_\perp; z) &= E_0\psi_G(\boldsymbol{r}_\perp; z)\left[\left(1 + \frac{x^2-y^2}{4q^2} + \frac{\alpha^2\bar{\lambda}^2}{2}\frac{x^2-y^2}{r_\perp^2}\right)J_0(\alpha r_\perp)\right. \\
&\quad \left. - \left(\alpha\bar{\lambda}^2\frac{x^2-y^2}{r_\perp^3} - i\frac{\alpha k\bar{\lambda}^2}{2q}\frac{x^2-y^2}{r_\perp} - i\frac{\alpha k\bar{\lambda}^2}{q}\frac{xy}{r_\perp}\right)J_1(\alpha r_\perp)\right] \\
cB_z(\boldsymbol{r}_\perp; z) &= E_0\psi_G(\boldsymbol{r}_\perp; z)\left(-\frac{y}{q}J_0(\alpha r_\perp) - i\alpha\bar{\lambda}\frac{y}{r}J_1(\alpha r_\perp)\right).
\end{aligned}
\tag{273}
$$

The Bessel-Gaussian Maxwell beams are generalizations of the Gaussian Maxwell and Bessel Maxwell beams respectively.

7.6 Cross polarization in Airy Maxwell beams

In 1979, Berry and Balázs demonstrated a nonspreading Airy wave packet solution to the Schrödinger equation (Berry & Balazs, 1979). The paraxial Helmholtz equation in light optics has the mathematical structure of the Schrödinger equation. Inspired by this quantum mechanical analog, the Airy 'light' beams were created in 2007 (Siviloglou et al., 2007). Since then, the Airy beam solutions of the paraxial Helmholtz equation have been drawing a lot of attention, as these beams have unusual and interesting properties (Mandel & Wolf, 1995; Otte, 2020). The novel features of Airy beams include, self-acceleration, self-healing and self-autofocusing (Broky et al., 2008; Driben et al., 2013; Hu et al., 2012; Rogel-Salazar et al., 2014; Siviloglou et al., 2007, 2008; Zhang et al., 2011). Airy beams have found applications in beam arrays and particle accelerators (Nomoto et al., 2015).

A scalar Airy beam is described by

$$\begin{aligned}\psi_A(\boldsymbol{r}_\perp; z) = {} & \mathrm{Ai}\left[\tilde{x} + ia\tilde{z} - \left(\frac{\tilde{z}}{2}\right)^2\right]\mathrm{Ai}\left[\tilde{y} + ia\tilde{z} - \left(\frac{\tilde{z}}{2}\right)^2\right] \\ & \times \exp\left[a\left(\tilde{x} - \frac{\tilde{z}^2}{2}\right) + i\left(\frac{\tilde{x}\tilde{z}}{2} + \frac{a^2\tilde{z}}{2} - \frac{\tilde{z}^3}{12}\right)\right] \\ & \times \exp\left[a\left(\tilde{y} - \frac{\tilde{z}^2}{2}\right) + i\left(\frac{\tilde{y}\tilde{z}}{2} + \frac{a^2\tilde{z}}{2} - \frac{\tilde{z}^3}{12}\right)\right],\end{aligned} \tag{274}$$

with $\tilde{x} = x/x_0$, $\tilde{y} = y/x_0$, and $\tilde{z} = \lambda\!\!\!^{-} z/x_0^2$ where x_0 is a scaling parameter to be determined experimentally (see Nomoto et al., 2015, for details). Ai $[X]$ denotes the Airy function of the argument X. To find the Airy Maxwell beam corresponding to the scalar beam in (274), we can follow the differential operator form of the MSS rule (234). As before we take the initial plane wave field distribution at the transverse plane at $z = 0$ as

$$\mathbb{F}(0) = E_0 \begin{pmatrix} 1 \\ 0 \\ 0 \\ 0 \\ 1 \\ 0 \end{pmatrix}. \tag{275}$$

This leads us to the field distribution for the Airy Maxwell beam as given in (235), namely,

$$
\begin{aligned}
E_x(\boldsymbol{r}_\perp; z) &= E_0\left[\psi_A(\boldsymbol{r}_\perp; z) + \frac{1}{4}\bar{\lambda}^2\left(\frac{\partial^2\psi_A(\boldsymbol{r}_\perp; z)}{\partial x^2} - \frac{\partial^2\psi_A(\boldsymbol{r}_\perp; z)}{\partial y^2}\right)\right], \\
E_y(\boldsymbol{r}_\perp; z) &= \frac{1}{2}\bar{\lambda}^2 E_0\left(\frac{\partial^2\psi_A(\boldsymbol{r}_\perp; z)}{\partial x \partial y}\right), \\
E_z(\boldsymbol{r}_\perp; z) &= i\bar{\lambda} E_0 \frac{\partial\psi_A(\boldsymbol{r}_\perp; z)}{\partial x}, \\
cB_x(\boldsymbol{r}_\perp; z) &= \frac{1}{2}\bar{\lambda}^2 E_0\left(\frac{\partial^2\psi_A(\boldsymbol{r}_\perp; z)}{\partial x \partial y}\right), \\
cB_y(\boldsymbol{r}_\perp; z) &= E_0\left[\psi_A(\boldsymbol{r}_\perp; z) - \frac{1}{4}\bar{\lambda}^2\left(\frac{\partial^2\psi_A(\boldsymbol{r}_\perp; z)}{\partial x^2} - \frac{\partial^2\psi_A(\boldsymbol{r}_\perp; z)}{\partial y^2}\right)\right], \\
cB_z(\boldsymbol{r}_\perp; z) &= i\bar{\lambda} E_0 \frac{\partial\psi_A(\boldsymbol{r}_\perp; z)}{\partial y}.
\end{aligned}
\tag{276}
$$

Airy beams are of relatively recent origin (Nomoto et al., 2015) and have now been generalized to anisotropic Airy beams (Khan, 2023c).

7.7 Cross polarization in anisotropic Airy Maxwell beams

A scalar anisotropic Airy beam is described by (Khan, 2023c)

$$
\begin{aligned}
\psi_{aA}(\boldsymbol{r}_\perp; z) = {} & \mathrm{Ai}\left[\tilde{x} + ia\tilde{z} - \left(\frac{\tilde{z}}{2}\right)^2\right]\mathrm{Ai}\left[\tilde{y} + ib\tilde{z} - \left(\frac{\tilde{z}}{2}\right)^2\right] \\
& \times \exp\left[a\left(\tilde{x} - \frac{\tilde{z}^2}{2}\right) + i\left(\frac{\tilde{x}\tilde{z}}{2} + \frac{a^2\tilde{z}}{2} - \frac{\tilde{z}^3}{12}\right)\right] \\
& \times \exp\left[b\left(\tilde{y} - \frac{\tilde{z}^2}{2}\right) + i\left(\frac{\tilde{y}\tilde{z}}{2} + \frac{b^2\tilde{z}}{2} - \frac{\tilde{z}^3}{12}\right)\right].
\end{aligned}
\tag{277}
$$

It is a generalized scalar Airy beam with unequal beam apertures in the x and y dimensions respectively. Hence, it does not have a rotational symmetry about the beam axis as in the case of the scalar isotropic Airy beams (Nomoto et al., 2015).

To get the anisotropic Airy Maxwell beam we proceed exactly as before. We take the initial plane wave field distribution at the transverse plane at $z = 0$ as

$$
\mathbb{F}(0) = E_0\begin{pmatrix} 1 \\ 0 \\ 0 \\ 0 \\ 1 \\ 0 \end{pmatrix},
\tag{278}
$$

and use the differential operator form of the MSS rule (234). This leads to the field distribution for the anisotropic Airy Maxwell beam as

$$
\begin{aligned}
E_x(\boldsymbol{r}_\perp; z) &= E_0\left[\psi_{aA}(\boldsymbol{r}_\perp; z) + \frac{1}{4}\bar{\lambda}^2\left(\frac{\partial^2\psi_{aA}(\boldsymbol{r}_\perp; z)}{\partial x^2} - \frac{\partial^2\psi_{aA}(\boldsymbol{r}_\perp; z)}{\partial y^2}\right)\right], \\
E_y(\boldsymbol{r}_\perp; z) &= \frac{1}{2}\bar{\lambda}^2 E_0\left(\frac{\partial^2\psi_{aA}(\boldsymbol{r}_\perp; z)}{\partial x\partial y}\right), \\
E_z(\boldsymbol{r}_\perp; z) &= i\bar{\lambda}E_0\frac{\partial\psi_{aA}(\boldsymbol{r}_\perp; z)}{\partial x}, \\
cB_x(\boldsymbol{r}_\perp; z) &= \frac{1}{2}\bar{\lambda}^2 E_0\left(\frac{\partial^2\psi_{aA}(\boldsymbol{r}_\perp; z)}{\partial x\partial y}\right), \\
cB_y(\boldsymbol{r}_\perp; z) &= E_0\left[\psi_{aA}(\boldsymbol{r}_\perp; z) - \frac{1}{4}\bar{\lambda}^2\left(\frac{\partial^2\psi_{aA}(\boldsymbol{r}_\perp; z)}{\partial x^2} - \frac{\partial^2\psi_{aA}(\boldsymbol{r}_\perp; z)}{\partial y^2}\right)\right], \\
cB_z(\boldsymbol{r}_\perp; z) &= i\bar{\lambda}E_0\frac{\partial\psi_{aA}(\boldsymbol{r}_\perp; z)}{\partial y}.
\end{aligned}
\tag{279}
$$

Airy beams are of relatively recent origin (Nomoto et al., 2015) and we have generalized them to anisotropic Airy beams (Khan, 2023c).

8. Concluding remarks

When we embarked on the journey to develop a matrix formulation of the Maxwell vector wave optics, we closely examined the wide variety of matrix representations spread over a century. To our surprise, none of the existing matrix representations were found to be adequate for developing a matrix formulation of the Maxwell vector wave optics! A possible reason for this is that the numerous matrix representations were derived for entirely different purposes (symmetries, comparison with the Dirac spinor equation and so on) and different approximations. Most of these representations were for a medium with constant refractive index! This motivated us to derive a new matrix representation *ab initio* for a linear inhomogeneous medium. Such a representation had been obtained heuristically by inspection of the equations obeyed by the Riemann-Silberstein-Weber (RSW) vector by one of us (see, Khan, 2005a). Few points about the new matrix representation are relevant. In the specially derived matrix representation, the RSW vector emerges as a natural basis, rather than being used as a starting building block. Maxwell equations for a source-free homogeneous medium written in a matrix form with $(\sqrt{\epsilon}E_x, \sqrt{\epsilon}E_y, \sqrt{\epsilon}E_z, 0, B_x/\sqrt{\mu}, B_y/\sqrt{\mu}, B_z/\sqrt{\mu}, 0)$ as the eight-dimensional basis vector makes it a Schrödinger-like equation with a Dirac-like matrix Hamiltonian which reduces to a direct sum of Pauli matrix blocks under a unitary transformation with the new basis vector expressed in terms of the RSW vector. Thus, the RSW vector emerges as a natural basis for a matrix representation of the Maxwell equations (Khan & Jagannathan, 2024c). We note, that this matrix representation of Maxwell equations can handle the electromagnetic fields, when expressed using a complex representation, which is often the case in studies on light beam optics and light polarization. The new matrix representation (Khan & Jagannathan, 2024c) has clarified several issues in the diverse set of matrix representations of the Maxwell equations. This novel matrix representation has served the chief purpose of developing a unified treatment of light beam optics and polarization. This matrix representation is used to derive an exact beam-optical Hamiltonian, which has the mathematical structure of the Dirac Hamiltonian in relativistic quantum mechanics. This enables us to study wave propagation problems using the routine techniques from quantum mechanics. In particular, we are able to apply the Foldy-Wouthuysen transformation technique to the beam optical Hamiltonian and expand it in a series of paraxial and aberrating terms order-by-order. Each order sub-Hamiltonian explains the aberrations of the corresponding order. This is analogous to the problem of expanding the Dirac Hamiltonian as non-relativistic part accompanied with relativistic corrections order-by-order. Consequently, the paraxial behavior along with aberrations to all orders are readily obtained and each of them gets modified by wavelength-dependent corrections.

In particular, we applied the formalism to two specific examples and saw how the scalar wave optics (paraxial behaviour and the aberrations) gets modified by the wavelength-dependent corrections. First we considered the example of medium with a constant refractive index, which is exactly solvable. This example enabled us to illustrate the approximation scheme and compared the infinite series results with the exact results. As a second example in light beam optics, we considered an axially symmetric graded-index medium. The Maxwell vector wave optical Hamiltonian for this system has wavelength-dependent correction terms in addition to the traditional Helmholtz scalar wave optical Hamiltonian. These wavelength-dependent corrections appear in two guises, with and without matrices. In the various traditional and non-traditional formalisms of the Helmholtz scalar wave optics there are only six third-order aberrations. From the experience of the charged particle optics of an axially symmetric magnetic lens, we know that the axial symmetry permits nine third-order aberrations. In our formalism, we get the remaining three aberrations, which are all purely wavelength-dependent. Thus our formalism modifies the six aberration coefficients of the Helmholtz optics and also give rise to the aforementioned three aberrations. Perhaps, the more interesting outcome is the image rotation proportional to the wavelength. We have derived an explicit expression for the angle in (181). Such, an image rotation has no analogue/counterpart in any of the scalar optics theories. It would be worthwhile to experimentally look for the predicted image rotation. The presence of the nine third-order aberrations and image rotation is well-known for an axially symmetric magnetic electron lens, even when treated classically. The quantum treatment of this system leads to the de Broglie wavelength-dependent corrections to the classical results (Jagannathan & Khan, 1996, 2019; Khan, 1997).

As for the light polarization, we had a detailed derivation of the Mukunda-Simon-Sudarshan matrix substitution rule (MSS rule) for transition from the Helmholtz scalar wave optics to the Maxwell vector wave optics. We further saw that the MSS rule can be written in terms of a three-term matrix differential operator. We further presented this matrix differential operator using 6×6 matrices, whose elements are simple partial derivatives (∂_x, ∂_y, ∂^2_{xx}, ∂^2_{yy} and $\partial^2_{xy} = \partial^2_{yx}$), that too only up to the second-order. This enabled us to develop a new and easier procedure for obtaining the electromagnetic field components in Maxwell optics, without resorting to the customary and difficult method of solving the Maxwell equations. Instead, our technique makes use of the readily available and easy to obtain solutions of the Helmholtz scalar equation. The electromagnetic field components were derived for a general Maxwell beam. In the complete set of the electromagnetic field components, we note, that there are certain characteristic relations among the components. In particular, we have $E_y = vB_x$ where E_y and B_x are the cross polarization components when the beam is x-polarized *i.e.*, for the beam E_x and B_y are the dominant electric and magnetic fields, respectively. Besides the cross polarization components there are also the longitudinal components E_z and B_z in the direction of propagation. Significantly, all beams with finite extent in the transverse directions have cross polarization and longitudinal components. Thus, our approach leads to new insights on the beam fields. The power of the newly developed differential operator technique for the application of the MSS rule has been demonstrated with examples of Bessel Maxwell, Bessel-Gaussian Maxwell, Airy Maxwell, and anisotropic Airy Maxwell beams. To date, the MSS substitution rule has been used only for the Gaussian Maxwell beams (Mukunda et al., 1983; Khan, 2021, 2023a; Mukunda et al., 1985a, 1985b; Simon et al., 1986, 1987), anisotropic Gaussian

Maxwell beams (Khan, 2024), Bessel Maxwell beams (Khan, 2022, 2023a) and the Bessel-Gaussian Maxwell beams (Khan, 2023d). In this Chapter we have applied the MSS substitution rule to the Hermite-Gaussian beams. The differential operator form of the MSS rule has been applied to the Airy beams (Khan, 2023c) and the anisotropic Airy beams (Khan, 2023c). The results obtained for the aforementioned beams are consistent with the results obtained by other methods. We hope that the matrix differential operator form of the MSS Substitution rule will be very useful as it translates the problem into computing second-order derivatives of the function governing the transverse profile of the beams. We have also covered the newly discovered anisotropic Airy beams (Khan, 2023c). Our formalism is a suitable candidate for extending the traditional theory of polarization beyond the paraxial approximation. This is so, because the paraxial limit of our formalism readily leads to the MSS rule for transition from the Helmholtz scalar wave optics to the Maxwell vector wave optics.

Our formalism of the Maxwell vector wave optics extends Hamilton's optico-mechanical analogy beyond the ray optics and scalar wave optics (Khan, 2017b; Korotkova & Testorf, 2023). Historical accounts of this analogy and its influence on quantum mechanics are to be found in (Forbes, 2001; Khan, 2007, 2014b, 2016e, 2016f, 2016g, 2017d).

Appendix A. Wave Equation for the electric field in an inhomogeneous medium

We shall now give the details of the derivation of the wave equation for $\boldsymbol{E}$ mentioned in (27) in Section 2. For the first component of $\mathcal{F}$, namely $\sqrt{\epsilon}\,E_x$ apart from the constant factor $1/\sqrt{2}$, we get from (24) and (26),

$$
\begin{aligned}
\frac{\partial^2(\sqrt{\epsilon}\,E_x)}{\partial t^2} = \; & v\{(-\partial_z + \partial_z\bar{\mu})\,v\,(-\partial_z + \partial_z\bar{\epsilon}) \\
& + (\partial_y - \partial_y\bar{\mu})\,v\,(\partial_y - \partial_y\bar{\epsilon}) \\
& + (-\partial_x - \partial_x\bar{\mu})\,v\,(-\partial_x - \partial_x\bar{\epsilon})\}(\sqrt{\epsilon}\,E_x) \\
& + v\{(\partial_y - \partial_y\bar{\mu})\,v\,(-\partial_x + \partial_x\bar{\epsilon}) \\
& + (-\partial_x - \partial_x\bar{\mu})\,v\,(-\partial_y - \partial_y\bar{\epsilon})\}(\sqrt{\epsilon}\,E_y) \\
& + v\{(-\partial_z + \partial_z\bar{\mu})\,v\,(\partial_x - \partial_x\bar{\epsilon}) \\
& + (-\partial_x - \partial_x\bar{\mu})\,v\,(-\partial_z - \partial_z\bar{\epsilon})\}(\sqrt{\epsilon}\,E_z).
\end{aligned}
\tag{A.1}
$$

Simplifying this equation leads us finally to

$$
\begin{aligned}
\frac{1}{v^2}\frac{\partial^2 E_x}{\partial t^2} = \; & \nabla^2 E_x + 2\Big\{\left(\partial^2_{xx}\bar{\epsilon}\right)E_x + \left(\partial^2_{xy}\bar{\epsilon}\right)E_y + \left(\partial^2_{xz}\bar{\epsilon}\right)E_z \\
& + (\partial_x\bar{\epsilon})\,\partial_x E_x + (\partial_y\bar{\epsilon})\,\partial_x E_y + (\partial_z\bar{\epsilon})\,\partial_x E_z\Big\} \\
& + \Big\{\partial_y\bar{\mu}\,(\partial_x E_y - \partial_y E_x) - \partial_z\bar{\mu}\,(\partial_z E_x - \partial_x E_z)\Big\} \\
= \; & \nabla^2 E_x + 2\frac{\partial}{\partial x}(\nabla\bar{\epsilon}\cdot\boldsymbol{E}) + 2(\nabla\bar{\mu}\times(\nabla\times\boldsymbol{E}))_x,
\end{aligned}
\tag{A.2}
$$

where $\partial^2_{xx} = \partial^2/\partial x^2$, $\partial^2_{xy} = \partial^2/\partial x\partial y = \partial^2/\partial y\partial x = \partial^2_{yx}$, and $\partial^2_{xz} = \partial^2/\partial x\partial z = \partial^2/\partial z\partial x = \partial^2_{zx}$. Working out the equations for the second and third components of $\mathcal{F}$ from (24) and (26) we obtain equations for E_y and E_z similar to (A.2) with x replaced by y and z, respectively. Thus, it is found that the resulting three equations are the three components of the equation

$$\nabla^2 \boldsymbol{E} - \epsilon\mu\frac{\partial^2 \boldsymbol{E}}{\partial t^2} = (\nabla\times \boldsymbol{E}) \times (\nabla \ln\mu) - \nabla\left((\nabla \ln\epsilon)\cdot\boldsymbol{E}\right), \tag{A.3}$$

which is the wave equation for $\boldsymbol{E}$ in an inhomogeneous medium (Born & Wolf, 1999). Using the well known vector calculus identities

$$\begin{aligned} \nabla\times(\nabla\varphi) &= 0, \\ \nabla(\boldsymbol{A}\cdot\boldsymbol{B}) &= (\boldsymbol{A}\cdot\nabla)\boldsymbol{B} + (\boldsymbol{B}\cdot\nabla)\boldsymbol{A} \\ &\quad + \boldsymbol{A}\times(\nabla\times\boldsymbol{B}) + \boldsymbol{B}\times(\nabla\times\boldsymbol{A}), \end{aligned} \tag{A.4}$$

where φ is a scalar function of $\boldsymbol{r}$ and $\boldsymbol{A}$ and $\boldsymbol{B}$ are vector functions of $\boldsymbol{r}$, we can write (A.3) as

$$\begin{aligned} \nabla^2\boldsymbol{E} - \epsilon\mu\tfrac{\partial^2 \boldsymbol{E}}{\partial t^2} &= (\nabla\times\boldsymbol{E})\times((\nabla\ln\epsilon)\cdot\nabla)\boldsymbol{E} \\ &\quad + (\boldsymbol{E}\cdot\nabla)(\nabla\ln\epsilon) + (\nabla(\ln\epsilon + \ln\mu)), \end{aligned} \tag{A.5}$$

another form of the generalized wave equation for $\boldsymbol{E}$ in an inhomogeneous medium (Mazharimousavi et al., 2013). It is to be noted that the equation for the fourth component, or the null component, of $\mathcal{F}$ leads to $0 = 0$ exactly. The equations for the fifth to seventh components of $\mathcal{F}$ lead to the wave equation for the magnetic field in an inhomogeneous medium (Born & Wolf, 1999). The equation for the eighth, or the null component, of $\mathcal{F}$, leads to $0 = 0$ exactly.

Appendix B. Riemann-Silberstein-Weber vector

Here we shall list some of the main properties of the Riemann-Silberstein-Weber (RSW) complex vector (Kiesslinga & Tahvildar-Zadehb, 2018; Sebens, 2019; Silberstein, 1907a, 1907b) appearing naturally in (32)–(33). Defined by

$$\boldsymbol{F}(\boldsymbol{r}, t) = \frac{1}{\sqrt{2}}\left(\sqrt{\epsilon(\boldsymbol{r})}\,\boldsymbol{E}(\boldsymbol{r}, t) + i\frac{1}{\sqrt{\mu(\boldsymbol{r})}}\boldsymbol{B}(\boldsymbol{r}, t)\right), \tag{B.1}$$

it has been called in the literature mostly as the Riemann-Silberstein vector. Following (Ram et al., 2021; Vahala et al., 2022a, 2022) we shall call it as the Riemann-Silberstein-Weber vector in view of the significance of the contributions to it by Weber (Kiesslinga & Tahvildar-Zadehb, 2018; Sebens, 2019). The RSW vector is a mixture of both a vector and a pseudovector. But it is to be noted that $(\boldsymbol{F}^+, \boldsymbol{F}^-)$ transforms according to the Lorentz group extended by parity (Wang, 2015).

Riemann-Silberstein-Weber vector can be used to express many quantities associated with the electromagnetic field. With $\boldsymbol{F}$ denoting the column vector of its components,

Poynting vector:

$$\boldsymbol{S} = \frac{1}{\mu}\boldsymbol{E}\times\boldsymbol{B} = -iv\left(\boldsymbol{F}^\dagger \times \boldsymbol{F}\right),$$

Energy density:

$$u = \frac{1}{2}\left(\epsilon \boldsymbol{E}\cdot\boldsymbol{E} + \frac{1}{\mu}\boldsymbol{B}\cdot\boldsymbol{B}\right) = \boldsymbol{F}^{\dagger}\cdot\boldsymbol{F},$$

Momentum density:

$$\boldsymbol{p} = \epsilon\,(\boldsymbol{E}\times\boldsymbol{B}) = -\frac{i}{v}(\boldsymbol{F}^{\dagger}\times\boldsymbol{F}),$$

Angular momentum density:

$$L = \epsilon\,\{\boldsymbol{r}\times(\boldsymbol{E}\times\boldsymbol{B})\} = -\frac{i}{v}\{\boldsymbol{r}\times(\boldsymbol{F}^{\dagger}\times\boldsymbol{F})\}, \tag{B.2}$$

where † is the Hermitian conjugate. The other quantities are:

Total energy:

$$\begin{aligned} E &= \tfrac{1}{2}\int_{-\infty}^{\infty}\int_{-\infty}^{\infty}\int_{-\infty}^{\infty} dx\,dy\,dz\left\{\epsilon\boldsymbol{E}\cdot\boldsymbol{E} + \tfrac{1}{\mu}\boldsymbol{B}\cdot\boldsymbol{B}\right\} \\ &= \int_{-\infty}^{\infty}\int_{-\infty}^{\infty}\int_{-\infty}^{\infty} dx\,dy\,dz\,\{\boldsymbol{F}^{\dagger}\cdot\boldsymbol{F}\}, \end{aligned}$$

Total momentum:

$$\begin{aligned} \boldsymbol{P} &= \int_{-\infty}^{\infty}\int_{-\infty}^{\infty}\int_{-\infty}^{\infty} dx\,dy\,dz\,\epsilon\,\{\boldsymbol{E}\times\boldsymbol{B}\} \\ &= -\tfrac{i}{v}\int_{-\infty}^{\infty}\int_{-\infty}^{\infty}\int_{-\infty}^{\infty} dx\,dy\,dz\,\{\boldsymbol{F}^{\dagger}\times\boldsymbol{F}\}, \end{aligned}$$

Total angular momentum:

$$\begin{aligned} \boldsymbol{M} &= \epsilon\int_{-\infty}^{\infty}\int_{-\infty}^{\infty}\int_{-\infty}^{\infty} dx\,dy\,dz\,\{\boldsymbol{r}\times(\boldsymbol{E}\times\boldsymbol{B})\} \\ &= -\tfrac{i}{v}\int_{-\infty}^{\infty}\int_{-\infty}^{\infty}\int_{-\infty}^{\infty} dx\,dy\,dz\,\{\boldsymbol{r}\times(\boldsymbol{F}^{\dagger}\times\boldsymbol{F})\}, \end{aligned}$$

Moment of energy:

$$\begin{aligned} \boldsymbol{N} &= \tfrac{1}{2}\int_{-\infty}^{\infty}\int_{-\infty}^{\infty}\int_{-\infty}^{\infty} dx\,dy\,dz\left\{\boldsymbol{r}\left(\epsilon\boldsymbol{E}\cdot\boldsymbol{E} + \tfrac{1}{\mu}\boldsymbol{B}\cdot\boldsymbol{B}\right)\right\} \\ &= \int_{-\infty}^{\infty}\int_{-\infty}^{\infty}\int_{-\infty}^{\infty} dx\,dy\,dz\,\{\boldsymbol{r}\,(\boldsymbol{F}^{\dagger}\cdot\boldsymbol{F})\}. \end{aligned} \tag{B.3}$$

In this form these quantities look like the quantum-mechanical expectation values!

It has been suggested (Bialynicki-Birula, 1994, 1996a, b) that $\boldsymbol{F}^{+}(\boldsymbol{r}, t)$, or $\boldsymbol{F}^{-}(\boldsymbol{r}, t)$, can be regarded as a three-component wave function of the photon. In this connection the following may be noted. With both $\sqrt{\epsilon}\boldsymbol{E}$ and $\boldsymbol{B}/\sqrt{\mu}$ in $\mathcal{F}$ having the dimension of square root of energy, we get

$$\mathcal{F}^{\dagger}(\boldsymbol{r}, t)\mathcal{F}(\boldsymbol{r}, t) = \frac{1}{2}\left(\epsilon|\boldsymbol{E}(\boldsymbol{r}, t)|^{2} + \frac{1}{\mu}|\boldsymbol{B}(\boldsymbol{r}, t)|^{2}\right), \tag{B.4}$$

the local energy density at the position $\boldsymbol{r}$ at time t. Now, from (32) we see that

$$\begin{aligned} \underline{\Psi}^{\dagger}\underline{\Psi} &= \mathcal{F}^{\dagger}\mathbb{T}^{\dagger}\mathbb{T}\mathcal{F} = \mathcal{F}^{\dagger}\mathcal{F} \\ &= \frac{1}{2}\left(\epsilon|\boldsymbol{E}|^{2} + \frac{1}{\mu}|\boldsymbol{B}|^{2}\right), \end{aligned} \tag{B.5}$$

where the elements of $\underline{\Psi}$ contain the components of $\boldsymbol{F}$. It is seen that the equation (B.5) is consistent with (B.2). If we let

$$\underline{\Psi}(\boldsymbol{r}, t) = \langle \boldsymbol{r}|\underline{\Psi}(t)\rangle, \tag{B.6}$$

then we have

$$\begin{aligned} \langle\underline{\Psi}(t)|\underline{\Psi}(t)\rangle &= \int_{-\infty}^{\infty}\int_{-\infty}^{\infty}\int_{-\infty}^{\infty} dx\, dy\, dz\, \underline{\Psi}^{\dagger}(\boldsymbol{r}, t)\underline{\Psi}(\boldsymbol{r}, t) \\ &= \int_{-\infty}^{\infty}\int_{-\infty}^{\infty}\int_{-\infty}^{\infty} dx\, dy\, dz\, \mathcal{F}^{\dagger}(\boldsymbol{r}, t)\mathcal{F}(\boldsymbol{r}, t) \\ &= \frac{1}{2}\int_{-\infty}^{\infty}\int_{-\infty}^{\infty}\int_{-\infty}^{\infty} dx\, dy\, dz\, \left(\epsilon|\boldsymbol{E}(\boldsymbol{r}, t)|^{2} + \frac{1}{\mu}|\boldsymbol{B}(\boldsymbol{r}, t)|^{2}\right), \end{aligned} \tag{B.7}$$

the total energy of the field at time t. If we write the time evolution equation (38) for $\underline{\Psi}(t)$ in a source-free homogeneous medium as

$$\frac{\partial|\underline{\Psi}(t)\rangle}{\partial t} = \nu\mathsf{M}_{0}\ |\underline{\Psi}(t)\rangle, \tag{B.8}$$

then, since $\nu\mathsf{M}_0$ is time-independent,

$$|\underline{\Psi}(t)\rangle = e^{\nu\mathsf{M}_{0}t}\ |\underline{\Psi}(0)\rangle, \qquad \langle\underline{\Psi}(t)| = \langle\underline{\Psi}(0)|\,(e^{\nu\mathsf{M}_{0}t})^{\dagger}. \tag{B.9}$$

The matrix M_0 in (38b) contains $\boldsymbol{\sigma}\cdot\nabla$ and $\boldsymbol{\sigma}^{*}\cdot\nabla$ as the diagonal blocks and zeros elsewhere. The three Pauli matrices $\{\boldsymbol{\sigma}\}$ are Hermitian. Then, since $\nabla^{\dagger} = -\nabla$, we have $\mathsf{M}_0^{\dagger} = -\mathsf{M}_0$. Thus, $(e^{\nu\mathsf{M}_{0}t})^{\dagger} = e^{-\nu\mathsf{M}_{0}t}$ and

$$\langle\underline{\Psi}(t)|\underline{\Psi}(t)\rangle = \langle\underline{\Psi}(0)|\underline{\Psi}(0)\rangle \tag{B.10}$$

showing that the total energy of the field is conserved in a source-free homogeneous medium.

The RSW vector can also be derived from the Hertz vector potential $\boldsymbol{Z}(\boldsymbol{r}, t)$, also known as the polarization potential (see, for example, Panofsky & Phillips, 1962):

$$\boldsymbol{F}(\boldsymbol{r}, t) = \nabla\times\left(\frac{i}{v}\frac{\partial\boldsymbol{Z}(\boldsymbol{r}, t)}{\partial t} + \nabla\times\boldsymbol{Z}(\boldsymbol{r}, t)\right). \tag{B.11}$$

The Hertz potential satisfies the wave equation

$$\nabla^{2}\boldsymbol{Z}(\boldsymbol{r}, t) - \frac{1}{v^{2}}\frac{\partial^{2}\boldsymbol{Z}(\boldsymbol{r}, t)}{\partial t^{2}} = 0. \tag{B.12}$$

Appendix C. Different formulations of ray optics and the Helmholtz scalar wave optics

In this appendix, first we summarise the Fermat-Hamilton formulation of ray optics. Then, after a brief comment on the Lie algebraic formulation of ray optics, we describe a non-traditional formulation of Helmholtz scalar wave optics.

C.1 Fermat-Hamilton formulation of ray optics

Historically, theoretical optics is based on Fermat's principle of least time (see, e.g., Buchdahl, 1993; Lakshminarayanan et al., 2002). Let light travel in a medium of refractive index $n(\boldsymbol{r})$ from a point P_{in} to a point P_{out}. If the direction of propagation of light is such that we can use the coordinate z to parametrize the path then the element of the path length ds along the path is

$$ds = \sqrt{dx^2 + dy^2 + dz^2} = \sqrt{(1 + x'^2 + y'^2)}\, dz, \tag{C.1}$$

where the prime $'$ denotes differentiation with respect to z. Let z_{in} and z_{out}, respectively, be the z coordinates of P_{in} and P_{out} which are the fixed initial and final points for the path of light. Fermat's principle of least time states that the optical path length along the path defined by

$$\begin{aligned} S &= \int_{P_{in}}^{P_{out}} ds\; n(\boldsymbol{r}(\boldsymbol{s})) \\ &= \int_{z_{in}}^{z_{out}} dz\; n(x(z), y(z), z)\sqrt{(1 + x'^2 + y'^2)}, \end{aligned} \tag{C.2}$$

is minimum for the actual path taken by light. This is best expressed in terms of the Euler-Lagrange equations

$$\frac{d}{dz}\left(\frac{\partial L}{\partial x'}\right) - \frac{\partial L}{\partial x} = 0, \qquad \frac{d}{dz}\left(\frac{\partial L}{\partial y'}\right) - \frac{\partial L}{\partial y} = 0, \tag{C.3}$$

where the ray optical, or the geometrical optical, Lagrangian $L(x, y, x', y'; z)$ is given by the expression

$$L = n(\boldsymbol{r})\sqrt{(1 + x'^2 + y'^2)}. \tag{C.4}$$

Using this Lagrangian, we obtain the following momenta, canonically conjugate to the transverse coordinates x and y by the standard prescription

$$\begin{aligned} \wp_x &= \frac{\partial L}{\partial x'} = n(\boldsymbol{r})\frac{x'}{\sqrt{1 + x'^2 + y'^2}} \\ \wp_y &= \frac{\partial L}{\partial y'} = n(\boldsymbol{r})\frac{y'}{\sqrt{1 + x'^2 + y'^2}}. \end{aligned} \tag{C.5}$$

Unlike in mechanics, these momenta are dimensionless quantities. In the present context, we need the ray optical Hamiltonian, which is obtained by the standard prescription

$$H = \wp_x x' + \wp_y y' - L = -\sqrt{n^2(\boldsymbol{r}) - \wp_\perp^2}. \tag{C.6}$$

The ray optical Lagrangian and Hamiltonian enable diverse techniques such as the eikonal methods, variational methods, ray tracing, which are well established in classical mechanics. The optical Lagrangian is the starting point to work out an alternate formalism adopting the techniques from the path integral formulation quantum mechanics (Robson et al., 2021). Significantly, the optical Hamiltonian leads to the Lie algebraic formulation of light beam optics (see Section 4).

The development of Lie methods applied to light beam optics is of a very recent origin. Same is true for the companion field of charged particle beam optics or particle accelerator optics. Both the fields have complemented each other in these new developments (Dragt, 1982, 1988; Dragt & Forest, 1986; Dragt et al., 1986, 1988; Lakshminarayanan et al., 1998; Rangarajan & Sachidanand, 1997; Rangarajan & Sridharan, 2010; Rangarajan et al., 1990). Two meetings on this emerging field of Lie methods applied to optics deserve a special mention. Both the meetings were held with the apt title 'Lie Methods in Optics' in Mexico and their proceedings have served a valuable resource to the practitioners of Lie methods in both light and charged particle beam optics (Khan, 2016h; Mondragón and Wolf, 1986; Wolf, 1989). In these meetings, it was pointed out that Lie methods complement the traditional methods of the eikonal and characteristic functions among others.

Prior to Lie methods, the aberration coefficients were obtained either by solving a trajectory equation by the method of variation of parameters (trajectory method, Green's function) or by differentiation of an appropriate eikonal function. The higher order aberrations are difficult to work out using such methods. In the Lie algebraic treatment, it is possible to calculate the aberration coefficients to much higher order with lesser efforts. Lie algebraic approach has clarified several issues in geometrical optics and has enabled a very precise way of enumerating and classifying aberrations in terms of polynomials derived from the Hamiltonian.

It is fascinating to note that one can arrive at Maxwell's equations starting from Fermat's principle of ray optics through a process of 'wavization', in analogy with the process of arriving at the wave equation of quantum mechanics from classical mechanics (Pradhan, 1987). Matrix methods have been used in other areas of optics and provide great advantages of simplifying and presenting the equations of optics (Gerrard & Burch, 1994). Same is true of other algebraic techniques in optics (Khan & Wolf, 2002).

C.2 A non-traditional formulation of the Helmholtz scalar wave optics

The transition from (82) to (83), in Section 4, reduces the original boundary value problem to a first order initial value problem in z. This reduction is of great practical value, since it leads to the powerful system of the Fourier optic approach and facilitates many applications (Goodman, 1996; Nazarathy & Shamir, 1980, 1982). However, the above reduction process is not found to be mathematically exact and rigorous enough! Hence, there is ample room to search for alternative procedures. It is seen that any reduction scheme is bound to be approximate and lack in rigor to some extent. The justification of a given reduction scheme lies primarily on the results it leads to. The occurrence of square-root operators in diverse situations have drawn a lot of interest (e.g., see Gill & Zachary, 2005, for details). Up to the paraxial approximation, there seems to be consensus among the known schemes pointing to the accuracy of paraxial wave optics. There have been significant differences, when going beyond the leading order paraxial approximation. We shall note some of these schemes.

It has been pointed out (Lax et al., 1975) that by neglecting $\boldsymbol{\nabla}(\boldsymbol{\nabla} \cdot \boldsymbol{E})$ in the derivation of the Helmholtz equation (see Appendix A) and seeking a solution that is plane polarized in the same sense everywhere is simply incompatible with the exact Maxwell equations. Using the angular spectrum representation for the electric field, the Gaussian beam propagation in a linear, homogeneous, and isotropic dielectric has been investigated and it has been demonstrated that the exact solution consists of the paraxial result plus higher-order non-Gaussian correction terms (Agrawal & Pattanayak, 1979). It has been further shown (Agrawal & Pattanayak, 1979) that this approach is consistent with the results of Lax et al. (1975). Significantly, Agarwal and Lax (1983) also examined the role of the boundary conditions in the corrections to the paraxial approximation of Gaussian beams. This work clarified several issues in other approaches, notably that of Couture and Belanger (1981) and Agrawal and Pattanayak (1979). Wünsche (1992) went a step further by constructing the linear transition operators and showed that this is equivalent to the complete integration of the system of coupled differential equations of Lax. Following this work of Wünsche, a simpler transition operator was derived (Cao, 1999; Cao & Deng, 1998). The transition operator method has been extended and used to generalize the uncertainty product to the case of non-paraxial fields (Alonso & Forbes, 2000; Alonso et al., 1999).

A non-traditional approach to the Helmholtz scalar wave optics was developed by exploiting the striking mathematical similarity between the Helmholtz equation and the Klein-Gordon equation of relativistic quantum mechanics. Thus, we obtained an alternative to the traditional square-root approach. From the very beginning the reader is reminded that the electromagnetic field and the Klein-Gordon field are two entirely different entities. But the underlying similarity in the mathematical structures of the two systems, enables us to use quantum methodology in Helmholtz optics. There was the suggestion to use the Foldy-Wouthuysen transformation in the case of the Helmholtz equation (Fishman & McCoy, 1984). In 1996, we independently chartered the same idea, while working out the quantum theory of charged particle beam optics (see, p.277 in Jagannathan & Khan, 1996). This idea was executed only in 2002 (Khan et al., 2002) and has since then developed into a complete formalism (Khan, 2005b, 2016b, 2018a, 2018c). The Foldy-Wouthuysen transformation has also found applications in acoustics (Fishman, 1992, 2004; Fishman et al., 2000).

At this stage, we depart from the traditional square-root procedure. We shall linearize (82) using a Feshbach-Villars-like procedure (Bjorken & Drell, 1994; Feshbach & Villars, 1958; Greiner, 2000; Jagannathan & Khan, 1996; Khan, 2017a). To this end, we introduce

$$\begin{pmatrix} \Psi_1(\boldsymbol{r}) \\ \Psi_2(\boldsymbol{r}) \end{pmatrix} = \begin{pmatrix} \Psi(\boldsymbol{r}) \\ -i\frac{\lambda\!\!\!^{-}}{n_c}\frac{\partial\Psi(\boldsymbol{r})}{\partial z} \end{pmatrix}. \tag{C.7}$$

The refractive index $n(\boldsymbol{r})$ is assumed to vary smoothly around a constant value n_c, such that $|n(\boldsymbol{r}) - n_c| \ll n_c$. Now, the equation (82) is written as

$$-i\frac{\lambda\!\!\!^{-}}{n_c}\frac{\partial}{\partial z}\begin{pmatrix} \Psi_1(\boldsymbol{r}) \\ \Psi_2(\boldsymbol{r}) \end{pmatrix} = \begin{pmatrix} 0 & 1 \\ \frac{1}{n_c^2}\left(n^2(\boldsymbol{r}) - \widehat{\wp}_\perp^2\right) & 0 \end{pmatrix}\begin{pmatrix} \Psi_1(\boldsymbol{r}) \\ \Psi_2(\boldsymbol{r}) \end{pmatrix}. \tag{C.8}$$

Equation (C.8) thus obtained has the structure of a first-order system. Next, we rewrite (C.8) in a Dirac-like form, using the following transform:

$$\begin{pmatrix} \Psi_1(\boldsymbol{r}) \\ \Psi_2(\boldsymbol{r}) \end{pmatrix} \longrightarrow \begin{pmatrix} \Psi_+(\boldsymbol{r}) \\ \Psi_-(\boldsymbol{r}) \end{pmatrix} = M\begin{pmatrix} \Psi_1(\boldsymbol{r}) \\ \Psi_2(\boldsymbol{r}) \end{pmatrix} = \frac{1}{\sqrt{2}}\begin{pmatrix} \Psi(\boldsymbol{r}) - i\frac{\lambda\!\!\!^{-}}{n_c}\frac{\partial\Psi(\boldsymbol{r})}{\partial z} \\ \Psi(\boldsymbol{r}) + i\frac{\lambda\!\!\!^{-}}{n_c}\frac{\partial\Psi(\boldsymbol{r})}{\partial z} \end{pmatrix}, \tag{C.9}$$

where

$$M = \frac{1}{\sqrt{2}}\begin{pmatrix} 1 & 1 \\ 1 & -1 \end{pmatrix} = M^{-1}. \tag{C.10}$$

The significance of the aforementioned transformation is evident, when we apply it to a monochromatic quasiparaxial beam propagating in the positive z-direction with the z-dependence

$$\Psi(\boldsymbol{r}) \sim \exp\left(\frac{i}{\lambda\!\!\!^{-}}n(\boldsymbol{r})z\right), \tag{C.11}$$

then

$$\begin{aligned}\Psi_+(\boldsymbol{r}) &\sim \tfrac{1}{\sqrt{2}}\left(1 + \tfrac{n(\boldsymbol{r})}{n_c}\right)\Psi(\boldsymbol{r}),\\ \Psi_-(\boldsymbol{r}) &\sim \tfrac{1}{\sqrt{2}}\left(1 - \tfrac{n(\boldsymbol{r})}{n_c}\right)\Psi(\boldsymbol{r}).\end{aligned} \tag{C.12}$$

we find that $\Psi_+(\boldsymbol{r})$ and $\Psi_-(\boldsymbol{r})$ satisfy the inequality $|\Psi_+(\boldsymbol{r})| \gg |\Psi_-(\boldsymbol{r})|$ enabling the application of the Foldy-Wouthuysen transformation technique. This helps us rewrite (C.9) readily in a Dirac-like form as

$$\begin{aligned} i\lambda\!\!\!^{-}\frac{\partial}{\partial z}\begin{pmatrix}\Psi_+(\boldsymbol{r})\\ \Psi_-(\boldsymbol{r})\end{pmatrix} &= \widehat{\mathrm{H}}\begin{pmatrix}\Psi_+(\boldsymbol{r})\\ \Psi_-(\boldsymbol{r})\end{pmatrix},\\ \widehat{\mathrm{H}} &= -n_c\sigma_z + \widehat{\mathcal{E}} + \widehat{\mathcal{O}},\\ \widehat{\mathcal{E}} &= \tfrac{1}{2n_c}\left\{\widehat{\wp}_\perp^2 + \left(n_c^2 - n^2(\boldsymbol{r})\right)\right\}\sigma_z,\\ \widehat{\mathcal{O}} &= \tfrac{i}{2n_c}\left\{\widehat{\wp}_\perp^2 + \left(n_c^2 - n^2(\boldsymbol{r})\right)\right\}\sigma_y,\end{aligned} \tag{C.13}$$

where $\widehat{\mathcal{E}}$ is an even operator and $\widehat{\mathcal{O}}$ is an odd operator. The beam optical Hamiltonian $\widehat{\mathrm{H}}$ is exact as is seen by squaring it

$$\widehat{\mathrm{H}}^2 = \left(n^2(\boldsymbol{r}) - \widehat{\wp}_\perp^2\right). \tag{C.14}$$

The first iteration in the FW-like transformation leads to the formal Hamiltonian to order $(\widehat{\wp}_\perp^2/n_c^2)^2$ as given by

$$\begin{aligned} i\lambda\!\!\!^{-}\frac{\partial}{\partial z}\begin{pmatrix}\Psi_+(\boldsymbol{r})\\ \Psi_-(\boldsymbol{r})\end{pmatrix} &= \widehat{H}^{(2)}\begin{pmatrix}\Psi_+(\boldsymbol{r})\\ \Psi_-(\boldsymbol{r})\end{pmatrix},\\ \widehat{H}^{(2)} &= -n_c\sigma_z + \widehat{\mathcal{E}} - \tfrac{1}{2n_c}\sigma_z\widehat{\mathcal{O}}^2.\end{aligned} \tag{C.15}$$

Considering that $|\Psi_+(\boldsymbol{r})| \gg |\Psi_-(\boldsymbol{r})|$ we can replace σ_z by 1_2. Then, identifying the dominant component $\Psi_+(\boldsymbol{r})$ with the Helmholtz scalar wave function $\Psi(\boldsymbol{r})$ we can write

$$i\lambda\!\!\!^{-}\frac{\partial\Psi(\boldsymbol{r})}{\partial z} = \widehat{H}^{(2)}\Psi(\boldsymbol{r}), \tag{C.16}$$

where the Hamiltonian takes the very familiar form in the traditional theory

$$\widehat{H}^{(2)} = -n_c + \frac{1}{2n_c}\left\{\widehat{\wp}_\perp^2 + \left(n_c^2 - n^2(\boldsymbol{r})\right)\right\} + \frac{1}{8n_c^3}\left\{\widehat{\wp}_\perp^2 + \left(n_c^2 - n^2(\boldsymbol{r})\right)\right\}^2. \tag{C.17}$$

The second iteration of the FW-like transformation leads to the formal Hamiltonian to order $(\widehat{\wp}_\perp^2/n_c^2)^4$ as given by

$$\begin{aligned} i\lambda\!\!\!^{-}\frac{\partial}{\partial z}\begin{pmatrix}\Psi_+(\boldsymbol{r})\\ \Psi_-(\boldsymbol{r})\end{pmatrix} &= \widehat{H}^{(4)}\begin{pmatrix}\Psi_+(\boldsymbol{r})\\ \Psi_-(\boldsymbol{r})\end{pmatrix},\\ \widehat{H}^{(4)} &= -n_c\sigma_z + \widehat{\mathcal{E}} - \tfrac{1}{2n_c}\sigma_z\widehat{\mathcal{O}}^2 - \tfrac{1}{8n_c^2}\left[\widehat{\mathcal{O}}, \left([\widehat{\mathcal{O}}, \widehat{\mathcal{E}}] + i\lambda\!\!\!^{-}\tfrac{\partial\widehat{\mathcal{O}}}{\partial z}\right)\right]\\ &\quad + \tfrac{1}{8n_c^3}\sigma_z\left\{\widehat{\mathcal{O}}^4 + \left([\widehat{\mathcal{O}}, \widehat{\mathcal{E}}] + i\lambda\!\!\!^{-}\tfrac{\partial\widehat{\mathcal{O}}}{\partial z}\right)^2\right\}.\end{aligned} \tag{C.18}$$

In the standard Foldy-Wouthuysen treatment of the Dirac Hamiltonian (see, e.g., Bjorken & Drell, 1994; Greiner, 2000), and also in the case of the Hamiltonian in the matrix formulation of the Maxwell vector wave optics (see Section 2), the odd term $\widehat{\mathcal{O}}$ is linear in $\widehat{\wp}_\perp$. So, recalling the theory of the Foldy-Wouthuysen-like transformations in Section 3, the first and second iterations of the FW-like transformation are seen to give terms up to $\widehat{\wp}_\perp^2$ and $\widehat{\wp}_\perp^4$, respectively. In (C.13), the odd term $\widehat{\mathcal{O}}$ is quadratic in $\widehat{\wp}_\perp$. So, the first and second iterations of the FW-like transformation give terms up to $\widehat{\wp}_\perp^4$ and $\widehat{\wp}_\perp^8$, respectively. After some algebra, replacing σ_z by 1_2, and identifying the dominant component $\Psi_+(\boldsymbol{r})$ with the Helmholtz scalar wave function $\Psi(\boldsymbol{r})$, as before, we can write

$$i\lambda\!\!\!^{-}\frac{\partial\Psi(\boldsymbol{r})}{\partial z} = \widehat{H}^{(4)}\Psi(\boldsymbol{r}) \tag{C.19}$$

where

$$\begin{aligned}\widehat{H}^{(4)} &= -n_c + \frac{1}{2n_c}\left\{\widehat{\wp}_\perp^2 + \left(n_c^2 - n^2(\boldsymbol{r})\right)\right\} + \frac{1}{8n_c^3}\left\{\widehat{\wp}_\perp^2 + \left(n_c^2 - n^2(\boldsymbol{r})\right)\right\}^2 \\ &\quad - \frac{i\lambda\!\!\!^{-}}{32n_c^4}\left[\widehat{\wp}_\perp^2, \frac{\partial(n^2(\boldsymbol{r}))}{\partial z}\right] + \frac{\lambda\!\!\!^{-2}}{32n_c^5}\left(\frac{\partial(n^2(\boldsymbol{r}))}{\partial z}\right)^2 + \frac{1}{16n_c^5}\left\{\widehat{\wp}_\perp^2 + \left(n_c^2 - n^2(\boldsymbol{r})\right)\right\}^3 \\ &\quad + \frac{5}{128n_c^7}\left\{\widehat{\wp}_\perp^2 + \left(n_c^2 - n^2(\boldsymbol{r})\right)\right\}^4.\end{aligned} \tag{C.20}$$

Before considering any specific system, we make a few remarks on the paraxial part of $\widehat{H}^{(4)}$, or the paraxial Hamiltonian,

$$\widehat{H}^{(p)} = -n_c + \frac{1}{2n_c}\left\{\widehat{\wp}_\perp^2 + \left(n_c^2 - n^2(\boldsymbol{r})\right)\right\} - \frac{i\lambda\!\!\!^{-}}{32n_c^4}\left[\widehat{\wp}_\perp^2, \frac{\partial(n^2(\boldsymbol{r}))}{\partial z}\right]. \tag{C.21}$$

The extra wavelength-dependent term in the paraxial Hamiltonian (C.21) is nonvanishing, whenever the refractive index of the medium possesses either longitudinal or transverse inhomogeneities. This term is real as the imaginary unit i present in it is nullified by the unit i coming from calculation of the commutator. This commutator term, $-(1/8n_c^2)[\widehat{\mathcal{O}}, ([\widehat{\mathcal{O}}, \widehat{\mathcal{E}}] + i\lambda\!\!\!^{-}\partial\widehat{\mathcal{O}}/\partial z)]$ originates from $\widehat{H}^{(4)}$ in (C.18). To get a better understanding, let us note that the corresponding term in the Foldy-Wouthuysen theory of the Dirac equation provides an accurate explanation of the spin-orbit term (including the Thomas precession effect) and the Darwin term attributed to the *zitterbewegung* (see, e.g., Bjorken & Drell, 1994; Greiner, 2000). The analogous term in the result of application of the Foldy-Wouthuysen transformation technique to the Feshbach-Villars form of the Klein-Gordon equation corresponds to the correction term to the classical electrostatic interaction of a point charge (see, e.g., Bjorken & Drell, 1994). In the present context, let us note that in the quantum theory of charged particle beam optics (based on the Klein-Gordon and Dirac equations respectively) the corresponding terms are quantum corrections to the classical paraxial and aberrating behaviours respectively. Guided by analogy, it would be interesting to study the influence of the commutator term on the propagation of Gaussian beams in a parabolic index medium, whose focusing strength is modulated in the axial variable z, however tiny it may be compared to the classical terms. It is to be noted that Hamilton's optico-mechanical analogy persists in the nontraditional formalism of the Helmholtz scalar wave optics (Khan, 2017b; Korotkova & Testorf, 2023).

Now, we shall examine the two systems we have seen in the Maxwell vector wave optics. These are medium with constant refractive index and an axially symmetric graded-index medium. It is to be noted that the impedance of the medium has played no role in the prescriptions for using the ray optics or the Helmholtz scalar wave optics. In ray optics, there is no impedance to start with. The derivation of the scalar wave equation from the Maxwell equations is an approximation and in the process the impedance gets eliminated (see Appendix A for details). Hence, the Helmholtz scalar wave optics does not consider the impedance of the medium. Impedance of the medium is present only in the matrix formulation of Maxwell vector wave optics. Its presence is recorded in the exact beam optical Hamiltonian (54) but later it may be omitted on certain physical grounds.

C.2.1 Medium with constant refractive index

For a medium with constant refractive index, $n(\boldsymbol{r}) = n_c$, from (C.13) we find the Hamiltonian to be

$$\begin{aligned} \widehat{\mathrm{H}}_c &= -n_c\sigma_z + \mathcal{E} + \mathcal{O} \\ &= -n_c\sigma_z + \hat{\mathfrak{t}}\sigma_z + \hat{\mathfrak{t}}(i\sigma_y), \qquad \hat{\mathfrak{t}} = \frac{\widehat{\wp}_\perp^2}{2n_c}. \end{aligned} \tag{C.22}$$

Note that σ_z is the β here. The Hamiltonian in (C.22) can be exactly diagonalized as follows. Following (105), let

$$\begin{aligned} \mathfrak{T} &= \exp\left(-\theta\sigma_z\mathcal{O}\right) = \exp\left(-\hat{\mathfrak{t}}\theta\sigma_x\right) \\ &= \cosh\left(\hat{\mathfrak{t}}\theta\right) - \sinh\left(\hat{\mathfrak{t}}\theta\right)\sigma_x, \end{aligned} \tag{C.23}$$

where we choose

$$\tanh\left(2\hat{\mathfrak{t}}\theta\right) = \frac{\hat{\mathfrak{t}}}{n_c - \hat{\mathfrak{t}}} = \frac{\widehat{\wp}_\perp^2}{2n_c^2 - \widehat{\wp}_\perp^2} < 1. \tag{C.24}$$

It follows that

$$\mathfrak{T}^{-1} = \exp\left(\hat{\mathfrak{t}}\theta\sigma_x\right) = \cosh\left(\hat{\mathfrak{t}}\theta\right) + \sinh\left(\hat{\mathfrak{t}}\theta\right)\sigma_x. \tag{C.25}$$

Then, we obtain

$$\begin{aligned} \widehat{\mathrm{H}}_c^{\text{diagonal}} &= \mathfrak{T}\widehat{\mathrm{H}}_c\mathfrak{T}^{-1} \\ &= -\sigma_z\sqrt{n_c^2 - 2n_c\hat{\mathfrak{t}}} = -\sigma_z\sqrt{n_c^2 - \widehat{\wp}_\perp^2}. \end{aligned} \tag{C.26}$$

Next, we compare the exact result thus obtained with the approximate one, obtained through the systematic series procedure we have developed. Let us define $P = \widehat{\wp}_\perp^2/n_c^2$. Then,

$$\begin{aligned} \widehat{H}_c^{(4)} &= -n_c\left(1 - \tfrac{1}{2}P - \tfrac{1}{8}P^2 - \tfrac{1}{16}P^3 - \tfrac{5}{128}P^4\right)\sigma_z \\ &\approx -n_c\sigma_z\sqrt{1 - P^2} \\ &= -\sigma_z\sqrt{n_c^2 - \widehat{\wp}_\perp^2} = \widehat{\mathrm{H}}_c^{\text{diagonal}}. \end{aligned} \tag{C.27}$$

C.2.2 Axially symmetric graded index medium

Our second example is that of an axially symmetric graded index medium. It is described by the axially symmetric refractive index profile

$$n\left(\boldsymbol{r}_{\perp}, z\right) = n_0 + \alpha_2(z) r_{\perp}^2 + \alpha_4(z) r_{\perp}^4 + \alpha_6(z) r_{\perp}^6 + \alpha_8(z) r_{\perp}^8 + \cdots . \tag{C.28}$$

Using the FW-like transformation technique described above one can get the beam optical Hamiltonian up to any desired order. We shall use the following notations to express the beam optical Hamiltonians compactly:

$$\begin{aligned}
\widehat{T} &= \left(\widehat{\wp}_{\perp} \cdot \boldsymbol{r}_{\perp} + \boldsymbol{r}_{\perp} \cdot \widehat{\wp}_{\perp}\right), \\
w_1(z) &= \frac{d}{dz}\left\{2 n_0 \alpha_2(z)\right\}, \\
w_2(z) &= \frac{d}{dz}\left\{\alpha_2^2(z) + 2 n_0 \alpha_4(z)\right\}, \\
w_3(z) &= \frac{d}{dz}\left\{2 n_0 \alpha_6(z) + 2 \alpha_2(z) \alpha_4(z)\right\}, \\
w_4(z) &= \frac{d}{dz}\left\{\alpha_4^2(z) + 2 \alpha_2(z) \alpha_6(z) + 2 n_0 \alpha_8(z)\right\}.
\end{aligned} \tag{C.29}$$

The paraxial and the aberrating Hamiltonians up to eighth order (governing seventh order aberrations) are

$$\begin{aligned}
\widehat{H} &= \widehat{H}_{0,p} + \widehat{H}_{0,(4)} + \widehat{H}_{0,(6)} + \widehat{H}_{0,(8)} + \widehat{H}_{0,(2)}^{(\bar{\lambda})} + \widehat{H}_{0,(4)}^{(\bar{\lambda})} + \widehat{H}_{0,(6)}^{(\bar{\lambda})} + \widehat{H}_{0,(8)}^{(\bar{\lambda})}, \\
\widehat{H}_{0,p} &= -n_0 + \frac{1}{2 n_0} \widehat{\wp}_{\perp}^2 - \alpha_2(z) r_{\perp}^2, \\
\widehat{H}_{0,(4)} &= \frac{1}{8 n_0^3} \widehat{\wp}_{\perp}^4 - \frac{\alpha_2(z)}{4 n_0^2}\left(\widehat{\wp}_{\perp}^2 r_{\perp}^2 + r_{\perp}^2 \widehat{\wp}_{\perp}^2\right) - \alpha_4(z) r_{\perp}^4, \\
\widehat{H}_{0,(6)} &= \frac{1}{16 n_0^5} \widehat{\wp}_{\perp}^6 - \frac{\alpha_2(z)}{8 n_0^4}\left\{\left(\widehat{\wp}_{\perp}^4 r_{\perp}^2 + r_{\perp}^2 \widehat{\wp}_{\perp}^4\right) + \widehat{\wp}_{\perp}^2 r_{\perp}^2 \widehat{\wp}_{\perp}^2\right\} \\
&\quad + \frac{1}{8 n_0^3}\left\{\left(\alpha_2^2(z) - 2 n_0 \alpha_4(z)\right)\left(\widehat{\wp}_{\perp}^2 r_{\perp}^4 + r_{\perp}^4 \widehat{\wp}_{\perp}^2\right) + 2 \alpha_2^2(z) r_{\perp}^2 \widehat{\wp}_{\perp}^2 r_{\perp}^2\right\} \\
&\quad - \alpha_6(z) r_{\perp}^6, \\
\widehat{H}_{0,(8)} &= \frac{5}{128 n_0^7} \widehat{\wp}_{\perp}^8 - \frac{5 \alpha_2(z)}{64 n_0^6}\left[\widehat{\wp}_{\perp}^4, \left[\widehat{\wp}_{\perp}^2, r_{\perp}^2\right]_{+}\right]_{+} \\
&\quad + \frac{1}{32 n_0^5}\left\{\left(3 \alpha_2^2(z) - 4 n_0 \alpha_4(z)\right)\left[\widehat{\wp}_{\perp}^4, r_{\perp}^4\right]_{+} + 5 \alpha_2^2(z)\left[\widehat{\wp}_{\perp}^2, r_{\perp}^2\right]_{+}^2\right. \\
&\quad \left. - \left(2 \alpha_2^2(z) + 4 n_0 \alpha_4(z)\right) \widehat{\wp}_{\perp}^2 r_{\perp}^4 \widehat{\wp}_{\perp}^2\right\} \\
&\quad + \frac{1}{16 n_0^4}\left\{4\left(\alpha_2^3(z) + n_0 \alpha_2(z) \alpha_4(z) + n_0^2 \alpha_6(z)\right)\left[\widehat{\wp}_{\perp}^2, r_{\perp}^6\right]_{+}\right. \\
&\quad - 5 \alpha_2^3(z)\left[r_{\perp}^4, \left[\widehat{\wp}_{\perp}^2, r_{\perp}^2\right]_{+}\right]_{+} \\
&\quad \left. + \left(2 \alpha_2^3(z) + 4 n_0 \alpha_2(z) \alpha_4(z)\right)\left[r_{\perp}^2, r_{\perp}^2 \widehat{\wp}_{\perp}^2 r_{\perp}^2\right]_{+}\right\} - \alpha_8(z) r_{\perp}^8, \\
\widehat{H}_{0\,(2)}^{(\bar{\lambda})} &= -\frac{\bar{\lambda}^2}{32 n_0^4} w_1(z) \widehat{T}, \\
\widehat{H}_{0,(4)}^{(\bar{\lambda})} &= -\frac{\bar{\lambda}^2}{32 n_0^4} w_2(z)\left(r_{\perp}^2 \widehat{T} + \widehat{T} r_{\perp}^2\right) + \frac{\bar{\lambda}^2}{32 n_0^5} w_1^2(z) r_{\perp}^4, \\
\widehat{H}_{0,(6)}^{(\bar{\lambda})} &= -\frac{3 \bar{\lambda}^2}{32 n_0^4} w_3(z)\left(r_{\perp}^4 \widehat{T} + \widehat{T} r_{\perp}^4\right) + \frac{\bar{\lambda}^2}{16 n_0^5} w_1(z) w_2(z) r_{\perp}^6, \\
\widehat{H}_{0,(8)}^{(\bar{\lambda})} &= -\frac{\bar{\lambda}^2}{8 n_0^4} w_4(z)\left(r_{\perp}^6 \widehat{T} + \widehat{T} r_{\perp}^6\right) + \frac{\bar{\lambda}^2}{32 n_0^5}\left\{w_2^2(z) + 2 w_1(z) w_3(z)\right\} r_{\perp}^8.
\end{aligned} \tag{C.30}$$

Starting with the Hamiltonian, up to fourth order, in (C.30) and using the same techniques used in the case of the Maxwell vector wave optics we obtain the six third order aberration coefficients:

$$
\begin{aligned}
C(z, z') &= \int_{z'}^{z} dz \left\{ \frac{1}{8n_0^3} D^4 - \frac{\alpha_2(z)}{2n_0^2} B^2D^2 - \alpha_4(z) B^4 \right. \\
&\quad \left. - \frac{\lambda\!\!\!^{-2}}{8n_0^4} w_2(z) B^3\, D + \frac{\lambda\!\!\!^{-2}}{32n_0^5} w_1^2(z) B^4 \right\}, \\
K(z, z') &= \int_{z'}^{z} dz \left\{ \frac{1}{8n_0^3} CD^3 - \frac{\alpha_2(z)}{4n_0^2} BD(AD + BC) - \alpha_4(z) AB^3 \right. \\
&\quad \left. - \frac{\lambda\!\!\!^{-2}}{32n_0^4} w_2(z) (B^2(AD + BC) + 2AB^2D) + \frac{\lambda\!\!\!^{-2}}{32n_0^5} w_1^2(z) AB^3 \right\}, \\
A(z, z') &= \int_{z'}^{z} dz \left\{ \frac{1}{8n_0^3} C^2D^2 - \frac{\alpha_2(z)}{2n_0^2}\, ABCD - \alpha_4(z) A^2B^2 \right. \\
&\quad \left. - \frac{\lambda\!\!\!^{-2}}{16n_0^4} w_2(z) (AB(AD + BC)) + \frac{\lambda\!\!\!^{-2}}{32n_0^5} w_1^2(z) A^2B^2 \right\}, \\
F(z, z') &= \int_{z'}^{z} dz \left\{ \frac{1}{8n_0^3} C^2D^2 - \frac{\alpha_2(z)}{4n_0^2} (A^2D^2 + B^2C^2) - \alpha_4(z) A^2B^2 \right. \\
&\quad \left. - \frac{\lambda\!\!\!^{-2}}{16n_0^4} w_2(z) (AB(AD + BC)) + \frac{\lambda\!\!\!^{-2}}{32n_0^5} w_1^2(z) A^2B^2 \right\}, \\
D(z, z') &= \int_{z'}^{z} dz \left\{ \frac{1}{8n_0^3} C^3\, D - \frac{\alpha_2(z)}{4n_0^2} AC(AD + BC) - \alpha_4(z) A^3B \right. \\
&\quad \left. - \frac{\lambda\!\!\!^{-2}}{32n_0^4} w_2(z) (A^2(AD + BC) + 2A^2BC) + \frac{\lambda\!\!\!^{-2}}{32n_0^5} w_1^2(z) A^3B \right\}, \\
E(z, z') &= \int_{z'}^{z} dz \left\{ \frac{1}{8n_0^3} C^4 - \frac{\alpha_2(z)}{2n_0^2} A^2C^2 - \alpha_4(z) A^4 - \frac{\lambda\!\!\!^{-2}}{8n_0^4} w_2(z) A^3\, C + \frac{\lambda\!\!\!^{-2}}{32n_0^5} w_1^2(z) A^4 \right\}.
\end{aligned}
\tag{C.31}
$$

Here, $A(z, z')$, $B(z, z')$, $C(z, z')$ and $D(z, z')$ are solutions of the paraxial ray equations, corresponding to the Hamiltonian started with, and satisfying (171)–(174). The third and higher order aberration coefficients are always expressed in terms of the solutions of the paraxial ray equations. This enables a better understanding of the aberration coefficients and serve as a guide in minimizing them (Dragt et al., 1986).

In the traditional Helmholtz scalar wave optics, one obtains six third-order aberrations. In the non-traditional treatment of the Helmholtz scalar wave optics, outlined above, all these six aberration coefficients are modified by wavelength-dependent corrections as seen above in (C.31) (Khan, 2005b, 2016b, 2018a, 2018c; Khan et al., 2002). Axial symmetry permits exactly nine third-order aberrations. Six of these are obtained in the traditional scalar wave theory and get modified by wavelength-dependent corrections in the non-traditional treatment of the Helmholtz scalar wave optics. As we have seen in Section 5, the scalar wave optics subduced by the Maxwell vector wave optics includes the remaining three third order aberrations as pure wavelength-dependent effects. We have obtained the expressions for all the nine aberration coefficients which explicitly have the wavelength-dependent corrections (see (201) in Section 5.2.3). An exclusive feature of the scalar wave optics subduced by the Maxwell vector wave optics is the presence of a wavelength-dependent angular momentum term. This leads to a small wavelength-dependent image rotation in an imaging system. We have derived its magnitude for a GRIN lens (see (181) in Section 5.2.2). This is analogous to the existence of nine aberrations and image rotation in the case of an axially symmetric magnetic lens in classical charged particle beam optics. The quantum treatment of this system modifies the classical aberration coefficients by the de Broglie wavelength-dependent correction terms (Jagannathan & Khan, 1996, 2019; Khan, 1997). The traditional Helmholtz scalar wave optics is obtained in the limit $\lambda\!\!\!^{-} \longrightarrow 0$.

Appendix D. Solutions of the Helmholtz equation

As we have seen earlier, the solutions of the Helmholtz scalar wave equation are used for deriving the cross polarization in various Maxwell beams. So, we shall have a closer look at the origins of these solutions of the scalar wave equation. We shall pay extra attention to the Airy and the anisotropic Airy beams as they are of relatively recent origin. We shall consider vacuum as the medium of propagation of light waves. For a monochromatic light wave the amplitude function $\Psi(\boldsymbol{r})$ will satisfy the Helmholtz equation (21):

$$\left[\left(\frac{\partial^2}{\partial x^2} + \frac{\partial^2}{\partial y^2} + \frac{\partial^2}{\partial z^2}\right) + k^2\right]\Psi(\boldsymbol{r}) = 0, \tag{D.1}$$

where $k = 1/\lambda\!\!\!^{-}$. For a quasiparaxial wave propagating predominantly along the z-direction we can write

$$\Psi(\boldsymbol{r}) = \psi(\boldsymbol{r})\exp(ikz). \tag{D.2}$$

Substituting this in (D.1), we obtain

$$\left(\frac{\partial^2}{\partial x^2} + \frac{\partial^2}{\partial y^2} + 2ik\frac{\partial}{\partial z} + \frac{\partial^2}{\partial z^2}\right)\psi(\boldsymbol{r}) = 0. \tag{D.3}$$

In the case of paraxial beams, the energy is primarily concentrated near the axis of the beam (chosen to be z-axis). Over propagation distances of the order of a few wavelengths, there is very little variation in the transverse profile of the beam governed by $\psi(\boldsymbol{r})$. Consequently, we have the following inequalities:

$$\begin{aligned}\frac{1}{k}\left|\frac{\partial\psi(\boldsymbol{r})}{\partial z}\right| &= \lambda\!\!\!^{-}\left|\frac{\partial\psi(\boldsymbol{r})}{\partial z}\right| \ll |\psi(\boldsymbol{r})|\\ \frac{1}{k}\left|\frac{\partial^2\psi(\boldsymbol{r})}{\partial z^2}\right| &= \lambda\!\!\!^{-}\left|\frac{\partial^2\psi(\boldsymbol{r})}{\partial z^2}\right| \ll \left|\frac{\partial\psi(\boldsymbol{r})}{\partial z}\right|.\end{aligned} \tag{D.4}$$

These lead us to approximate (D.3) as

$$\left(\frac{\partial^2}{\partial x^2} + \frac{\partial^2}{\partial y^2} + 2ik\frac{\partial}{\partial z}\right)\psi(\boldsymbol{r}) = 0, \tag{D.5}$$

which is known as the paraxial Helmholtz wave equation, or simply the paraxial wave equation. This paraxial wave equation has the mathematical structure of the time-dependent Schrödinger equation. It relates the variation of the beam profile $\psi(\boldsymbol{r})$ in the transverse plane (xy-plane) and the z-direction, which is the direction of propagation. Paraxial wave equation is known to have several families of solutions (Lakshminarayanan & Varadharajan, 2015).

D.1 Gaussian beams

Gaussian beams are the exact solutions of the paraxial wave equation (Kogelink & Li, 1966) and mathematically the simplest. The transformation properties of the Gaussian beams are completely described by the Kogelnik's ABCD law (Kogelnik, 1965). Gaussian beams continue to receive significant interest owing to the fact that they form the fundamental modes of laser cavities bounded by spherical mirrors. So, we shall first look at the Gaussian solutions. Let us consider a Gaussian beam propagating along the z-axis with its waist positioned at the transverse plane at $z = 0$ (without any loss of generality). The field in the waist plane is

$$\psi_G(\mathbf{r}_\perp; 0) = \sqrt{\frac{2}{\pi}}\frac{1}{\sigma_0}\exp\left(-\frac{1}{\sigma_0^2}r_\perp^2\right). \tag{D.6}$$

After propagation through a distance z along the z-direction the Gaussian beam has the transverse profile

$$\psi_G(\mathbf{r}_\perp; z) = \sqrt{\frac{2}{\pi}}\frac{1}{\sigma(z)}\exp\left(\frac{ik}{2q(z)}r_\perp^2\right). \tag{D.7}$$

The complex radius of curvature $q(z)$ is related to the radius of curvature of the phase front $R(z)$ as

$$\frac{1}{q(z)} = \frac{1}{R(z)} + \frac{2i}{k\sigma^2(z)}, \tag{D.8}$$

where $\sigma(z)$ is the beam width.

D.2 Anisotropic Gaussian beams

In addition to the Gaussian beam solutions, the paraxial wave equation also permits another class of exact solutions called as anisotropic Gaussian beam solutions (Simon, 1983, 1985, 1987). They form the resonant modes of the Fabry-Perot laser resonators with astigmatic mirrors and are of immense interest (Seshadri, 2009). The field of the anisotropic Gaussian beam is

$$\psi_{aG}(\mathbf{r}_\perp; z) = \sqrt{\frac{2}{\pi}}\frac{1}{\sqrt{\sigma_1(z)\sigma_2(z)}}\exp\left(\frac{ik}{2q_1(z)}x^2 + \frac{ik}{2q_2(z)}y^2\right). \tag{D.9}$$

The complex radii of curvature, $q_1(z)$ and $q_2(z)$, and the radii of curvature of the phase fronts, $R_1(z)$ and $R_2(z)$ are, respectively, related as

$$\begin{aligned}\frac{1}{q_1(z)} &= \frac{1}{R_1(z)} + \frac{2i}{k\sigma_1^2(z)},\\ \frac{1}{q_2(z)} &= \frac{1}{R_2(z)} + \frac{2i}{k\sigma_2^2(z)},\end{aligned} \tag{D.10}$$

where $\sigma_1(z)$ and $\sigma_2(z)$ are the unequal beam widths along two mutually perpendicular directions orthogonal to the beam axis (chosen to be z-axis).

D.3 Hermite-Gaussian beams

Another class of solutions is the Hermite-Gaussian beams (Erikson & Singh, 1994). It is to be noted that the laser resonators bounded by spherical mirrors have Gaussian beams as their fundamental modes and Hermite-Gaussian beams as higher-order modes. Hermite-Gaussian beams are a family of structurally stable laser modes, which have rectangular symmetry along the propagation axis. An interesting property of these beams is that they propagate in a shape-invariant manner, not only under free propagation but also through any first-order systems (Simon & Mukunda, 1998). Such beams play a fundamental role in the first-order optics, enabling the use of certain group theoretic approaches (Simon & Mukunda, 1998).

Hence, the Hermite-Gaussian beams have drawn immense interest. The Hermite-Gaussian beam is given by

$$\psi_{HG}(\boldsymbol{r}_\perp; z) = \sqrt{\frac{2}{\pi}}\frac{1}{\sigma(z)}H_m\left(\sqrt{2}\frac{x}{\sigma(z)}\right)H_n\left(\sqrt{2}\frac{y}{\sigma(z)}\right)\exp\left(\frac{ik}{2q(z)}\boldsymbol{r}_\perp^2\right), \quad \text{(D.11)}$$

where $H_m(x)$ and $H_n(y)$ are Hermite polynomials of degree m and n respectively (Conry et al., 2012; Erikson & Singh, 1994; Saghafi et al., 2001; Simon & Mukunda, 1998). The beam spot size in the plane that intersects the beam axis at z is given by $\sigma(z)$.

D.4 Bessel beams

It is known that the Helmholtz equation permits exact solutions that can propagate as beams without broadening and without divergence. These solutions are the Bessel functions. The solution of the lowest order has the transverse field distribution given by the $J_0\,(r_\perp;\, z)$, the zeroth order Bessel function. But $J_0\,(r_\perp;\, z)$ is not square integrable, implying that the beam has an infinite amount of energy! However, the Bessel beams have an academic interest (Wolf, 1988). Simplest Bessel beams have the form

$$\psi_B(\boldsymbol{r}_\perp; z) = J_0(\alpha r_\perp)\, e^{i\beta z}, \quad \text{(D.12)}$$

with $r_\perp = \sqrt{x^2 + y^2}$ (see Mishra, 1991; Sheppard et al., 2009).

D.5 Bessel-Gaussian beams

Mathematically, the simplest Bessel beams are the ones, whose transverse field distribution is the zeroth order Bessel function, which is not square integrable. Such beams are not physical, as they would require an infinite amount of energy to produce them. These nonphysical beams have drawn substantial academic interest (Mishra, 1991; Sheppard et al., 2009; Wolf, 1988). There have been attempts to create modified Bessel beams. Such beams carry finite power and are almost nondiffracting, as they can propagate over a long range without significant divergence. One such type is the Bessel-Gaussian beam (also called as Bessel-Gauss beam). The field amplitude distribution of a simplest Bessel-Gaussian beam can be expressed in the form (Li et al., 2004)

$$\psi_{BG}(\boldsymbol{r}_\perp; z) = \sqrt{\frac{2}{\pi}}\frac{1}{\sigma(z)}J_0(\alpha r_\perp)\exp\left(\frac{ik}{2q(z)}r_\perp^2\right). \quad \text{(D.13)}$$

D.6 Airy beams

We shall closely follow Nomoto et al. (2015) in describing the Airy beam solutions. Using the dimensionless and scaled variables, $\tilde{x} = x/x_0$, $\tilde{y} = y/x_0$ and $\tilde{z} = z/kx_0^2$, we write (D.5) as

$$\left(\frac{\partial^2}{\partial\tilde{x}^2} + \frac{\partial^2}{\partial\tilde{y}^2} + 2i\frac{\partial}{\partial\tilde{z}}\right)\psi(\boldsymbol{r}) = 0. \quad \text{(D.14)}$$

In order to obtain the Airy beam solutions, we shall look for solutions separable in $\tilde{x}$ and $\tilde{y}$ having the structure

$$\psi(\boldsymbol{r}) = u_x(\tilde{x}, \tilde{z})\, u_y(\tilde{y}, \tilde{z}). \quad \text{(D.15)}$$

Then, $u_x(\tilde{x}, \tilde{z})$ and $u_y(\tilde{y}, \tilde{z})$ are necessarily the solutions of the following pair of equations:

$$\begin{aligned}\left(\frac{\partial^2}{\partial \tilde{x}^2} + 2i\frac{\partial}{\partial \tilde{z}}\right) u_x(\tilde{x}, \tilde{z}) &= 0,\\ \left(\frac{\partial^2}{\partial \tilde{y}^2} + 2i\frac{\partial}{\partial \tilde{z}}\right) u_y(\tilde{y}, \tilde{z}) &= 0.\end{aligned} \tag{D.16}$$

Among the various solutions of these differential equations, we are interested in the Airy function solutions. The Airy functions have several definitions (Lakshminarayanan & Varadharajan, 2015) and in the present context, we note the following

$$\mathrm{Ai}(x) = \int_{-\infty}^{\infty} e^{ix\zeta + \frac{i}{3}\zeta^3} d\zeta, \tag{D.17}$$

where ζ is a complex variable. The Airy function defined by (D.17) satisfies the paraxial wave equation in (D.14). But such a solution is not physical, as it would require an infinite amount of energy to create such beams. In order to keep the solutions bounded and finite in energy, Nomoto et al. (2015) introduced a dimensionless factor in the form of an exponential function multiplying the Airy function. This exponential factor ensures that the resulting solutions approach zero, rather than infinitely oscillate along the x or y axis. The resulting solutions with the initial conditions (in the plane $z = 0$) are

$$\begin{aligned}u_x(\tilde{x}, 0) &= \mathrm{Ai}[\tilde{x}]e^{a\tilde{x}},\\ u_y(\tilde{y}, 0) &= \mathrm{Ai}[\tilde{y}]e^{a\tilde{y}},\end{aligned} \tag{D.18}$$

where $\mathrm{Ai}(\tilde{x})$ and $\mathrm{Ai}(\tilde{y})$ denote the Airy functions with the arguments $\tilde{x}$ and $\tilde{y}$ respectively. The parameter a corresponds to the effective aperture in the x and y directions respectively. The parameter a is confined to the range of 0–1. The parameter a truncates the Airy beam for negative values of $\tilde{x}$ and $\tilde{y}$. For $a = 0$, it leads to the ideal or infinite Airy beam and for $a \approx 1$ it approaches a single-lobe beam. Just like the infinite plane beam and the infinite Bessel beam, an infinite Airy beam is also unrealistic. Such beams would require an infinite amount of energy to produce them. The introduction of the exponential factor is fully justified through a Huygens-Fresnel integral (see Nomoto et al., 2015) and that process leads to the required solution

$$\begin{aligned}\psi_A(\mathbf{r}) &= \mathrm{Ai}\left[\tilde{x} + ia\tilde{z} - \left(\frac{\tilde{z}}{2}\right)^2\right]\mathrm{Ai}\left[\tilde{y} + ia\tilde{z} - \left(\frac{\tilde{z}}{2}\right)^2\right]\\ &\times \exp\left[a\left(\tilde{x} - \frac{\tilde{z}^2}{2}\right) + i\left(\frac{\tilde{x}\tilde{z}}{2} + \frac{a^2\tilde{z}}{2} - \frac{\tilde{z}^3}{12}\right)\right]\\ &\times \exp\left[a\left(\tilde{y} - \frac{\tilde{z}^2}{2}\right) + i\left(\frac{\tilde{y}\tilde{z}}{2} + \frac{a^2\tilde{z}}{2} - \frac{\tilde{z}^3}{12}\right)\right].\end{aligned} \tag{D.19}$$

The solution in (D.19) describes an Airy beam.

D.7 Anisotropic Airy beams

The anisotropic Airy beam solutions are obtained by modifying the equation (D.18) as

$$\begin{aligned}u_x(\tilde{x}, 0) &= \mathrm{Ai}[\tilde{x}]e^{a\tilde{x}},\\ u_y(\tilde{y}, 0) &= \mathrm{Ai}[\tilde{y}]e^{b\tilde{y}}.\end{aligned} \tag{D.20}$$

The parameters a and b correspond to the effective apertures in the x and y directions respectively. Both parameters a and b are confined to the range of 0–1. The parameters a and b truncate the Airy beam for negative values of $\tilde{x}$ and $\tilde{y}$ respectively. For $a = b = 0$, it leads to the ideal or infinite Airy beam and for $a = b \approx 1$ it approaches a single-lobe beam. The required solution is

$$\begin{aligned} \psi_{aA}(\boldsymbol{r}) = & \operatorname{Ai}\left[\tilde{x} + ia\tilde{z} - \left(\frac{\tilde{z}}{2}\right)^2\right] \operatorname{Ai}\left[\tilde{y} + ib\tilde{z} - \left(\frac{\tilde{z}}{2}\right)^2\right] \\ & \times \exp\left[a\left(\tilde{x} - \frac{\tilde{z}^2}{2}\right) + i\left(\frac{\tilde{x}\tilde{z}}{2} + \frac{a^2\tilde{z}}{2} - \frac{\tilde{z}^3}{12}\right)\right] \\ & \times \exp\left[b\left(\tilde{y} - \frac{\tilde{z}^2}{2}\right) + i\left(\frac{\tilde{y}\tilde{z}}{2} + \frac{b^2\tilde{z}}{2} - \frac{\tilde{z}^3}{12}\right)\right]. \end{aligned} \tag{D.21}$$

The solution in (D.21) describes an Airy beam with unequal beam apertures in the x and y dimensions respectively. Hence, it does not have a rotational symmetry about the beam axis followed by the isotropic Airy beams. The beam defined by (D.21) is referred to as anisotropic Airy beam (see Khan, 2023c). When $a = b$ the anisotropic Airy beam becomes an isotropic Airy beam (Nomoto et al., 2015).

References

Abramowitz, M., & Stegun, I. A. (2014). *Handbook of mathematical functions with formulas, graphs, and mathematical tables*. Dover. http://people.math.sfu.ca/~cbm/aands/; http://dlmf.nist.gov/.

Acharya, R., & Sudarshan, E. C. G. (1960). Front description in relativistic quantum mechanics. *Journal of Mathematical Physics, 1*, 532–536. https://doi.org/10.1063/1.1703689.

Agarwal, G. P., & Lax, M. (1983). Free-space wave propagation beyond the paraxial approximation. *Physical Review A, 27*, 1693–1695. https://doi.org/10.1103/PhysRevA.27.1693.

Agrawal, G. P., & Pattanayak, D. N. (1979). Gaussian beam propagation beyond the paraxial approximation. *Journal of the Optical Society of America A, 69*, 575–578. https://doi.org/10.1364/JOSA.69.000575.

Alonso, M. A., & Forbes, G. W. (2000). Uncertainty products for nonparaxial wave fields. *Journal of the Optical Society of America, 17*(12), 2391–2402. https://doi.org/10.1364/JOSAA.17.002391.

Alonso, MA., Asatryan, A. A., & Forbes, G. W. (1999). Beyond the Fresnel approximation for focused waves. *Journal of the Optical Society of America, 16*, 1958–1969. https://doi.org/10.1364/JOSAA.16.001958.

Arfken, G. B., Weber, H. J., & Harris, F. E. (2012). *Mathematical methods for physicists: A comprehensive guide* (7th ed.). Academic Press.

Barnett, S. M. (2014). Optical Dirac equation. *New Journal of Physics, 16*, 093008. https://doi.org/10.1088/1367-2630/16/9/093008.

Belkovich, I. V., & Kogan, B. L. (2016). Utilization of Riemann-Silberstein vectors in electromagnetics. *Progress in Electromagnetics Research B, 69*, 103–116.

Berry, M. V., & Balazs, N. L. (1979). Nonspreading wave packets. *American Journal of Physics, 47*, 264–267. https://doi.org/10.1119/1.11855.

Bialynicki-Birula, I. (1994). On the wave function of the photon. *Acta Physica Polonica A, 86*, 97–116. http://przyrbwn.icm.edu.pl/APP/ABSTR/86/a86-1-8.html.

Bialynicki-Birula, I. (1996a). The photon wave function. In H. H. Eberly, L. Mandel, & E. Wolf (Eds.). *Coherence and quantum optics VII* (pp. 313–322). Plenum Press. https://doi.org/10.1007/978-1-4757-9742-8_38.

Bialynicki-Birula, I. (1996b). The photon wave function. In E. Wolf (Vol. Ed.), *Progress in optics. Vol. XXXVI.* (pp. 245–294). Elsevier. https://doi.org/10.1016/S0079-6638(08)70316-0.

Bialynicki-Birula, I., & Bialynicka-Birula, Z. (2013). The role of the Riemann-Silberstein vector in classical and quantum theories of electromagnetism. *Journal of Physics A: Mathematical and Theoretical, 46*, 053001. https://doi.org/10.1088/1751-8113/46/5/053001.

Bjorken, J. D., & Drell, S. D. (1994). *Relativistic quantum mechanics*. McGraw-Hill.

Blanes, S., Casas, F., Oteo, J. A., & Ros, J. (2009). The Magnus expansion and some of its applications. *Physics Reports, 470*, 151–238.

Bocker, R. P., & Frieden, B. R. (1993). Solution of the Maxwell field equations in vacuum for arbitrary charge and current distributions using the methods of matrix algebra. *IEEE Transactions on Education, 36*, 350–356.

Bocker, R. P., & Frieden, B. R. (2018). A new matrix formulation of the Maxwell and Dirac equations. *Heliyon, 4*, e01033. https://doi.org/10.1016/j.heliyon.2018.e01033.

Bogush, A., Red'kov, V., Tokarevskaya, N., & Spix, G. (2009). Majorana-Oppenheimer approach to Maxwell electrodynamics in Riemannian space-time. arXiv:0905.0261[math-ph]. https://doi.org/10.48550/arXiv.0905.0261

Born, M., & Wolf, E. (1999). *Principles of optics*. Cambridge University Press.

Broky, J., Siviloglou, G., Dogariu, A., & Christodoulides, D. N. (2008). Self-healing properties of optical Airy beams. *Optics Express, 16*, 12880–12891. https://doi.org/10.1364/OE.16.012880.

Buchdahl, H. A. (1993). *An introduction to Hamiltonian optics*. Dover.

Cao, Q. (1999). Corrections to the paraxial approximation solutions in transversely non-uniform refractive-index media. *Journal of the Optical Society of America, 16*, 2494–2499. https://doi.org/10.1364/JOSAA.16.002494.

Cao, Q., & Deng, X. (1998). Corrections to the paraxial approximation of an arbitrary free-propagation beam. *Journal of the Optical Society of America, 15*, 1144–1148. https://doi.org/10.1364/JOSAA.15.001144.

Case, K. M. (1954). Some generalizations of the Foldy-Wouthuysen transformation. *Physical Review, 95*, 1323–1328. https://doi.org/10.1103/PhysRev.95.1323.

Chávez-Cerda, S. (1999). A new approach to bessel beams. *Journal of Modern Optics, 46*, 923–930. https://doi.org/10.1080/09500349908231313.

Conte, M., Jagannathan, R., Khan, S. A., & Pusterla, M. (1996). Beam optics of the Dirac particle with anomalous magnetic moment. *Particle Accelerators, 56*, 99–126. https://cds.cern.ch/record/307931/files/p99.pdf.

Conry, J. C., Vyas, R., & Singh, S. (2012). Cross-polarization of linearly polarized Hermite-Gauss laser beams. *Journal of the Optical Society of America, 29*, 579–584. https://doi.org/10.1364/JOSAA.29.000579.

Costella, J. P., & McKellar, B. H. J. (1995). The Foldy-Wouthuysen transformation. *American Journal of Physics, 63*, 1119–1121. https://doi.org/10.1119/1.18017.

Couture, M., & Belanger, P. A. (1981). From Gaussian beam to complex-source-point spherical wave. *Physical Review A, 24*, 355–359. https://doi.org/10.1103/PhysRevA.24.355.

Dragt, A. J. (1982). A Lie algebraic theory of geometrical optics and optical aberrations. *Journal of the Optical Society of America A, 72*, 372–379. https://doi.org/10.1364/JOSA.72.000372.

Dragt A. J. (1988). Lie algebraic method for ray and wave optics. University of Maryland. Department of Physics.

Dragt, A. J., & Forest, E. (1986). Lie algebraic theory of charged particle optics and electron microscopes. In P. W. Hawkes (Vol. Ed.), *Advances in imaging and electron physics. Vol. 67.* (pp. 65–120). Academic Press. https://doi.org/10.1016/S0065-2539(08)60329-7.

Dragt, A. J., Forest, E., & Wolf, K. B. (1986). Foundations of a Lie algebraic theory of geometrical optics. In J. S. Mondragón, & K. B. Wolf (Vol. Eds.), *Lie methods in optics. Lecture notes in physics. Vol. 250.* (pp. 105–157). Springer. https://doi.org/10.1007/3-540-16471-5_4.

Dragt, A. J., Neri, F., Rangarajan, G., Douglas, D. R., Healy, L. M., & Ryne, R. D. (1988). Lie algebraic treatment of linear and nonlinear beam dynamics. *Annual Review of Nuclear and Particle Science, 38*, 455–496. https://doi.org/10.1146/annurev.ns.38.120188.002323.

Driben, R., Hu, Y., Chen, Z., Malomed, B., & Morandotti, R. (2013). Inversion and tight focusing of Airy pulses under the action of third-order dispersion. *Optics Letters, 38*, 2499–2501. https://doi.org/10.1364/OL.38.002499.

Dvoeglazov, V. V. (1993). Electrodynamics with Weinberg's photons. *Hadronic Journal, 16*, 423–428. http://arxiv.org/abs/hep-th/9306108 arXiv:hep-th/9306108.

Edmonds, J. D. (1975). Comment on the Dirac-like equation for the photon. *Lettere al Nuovo Cimento, 13*, 185–186.

Erikson, W. L., & Singh, S. (1994). Polarization properties of Maxwell-Gaussian laser beams. *Physical Review E, 49*, 5778–5786. https://doi.org/10.1103/PhysRevE.49.5778.

Esposito, S. (1998). Covariant Majorana formulation of electrodynamics. *Foundations of Physics, 28*, 231–244. https://doi.org/10.1023/A:1018752803368.

Fedele, R., Miele, G., Palumbo, L., & Vaccaro, V. G. (1993). Thermal wave model for nonlinear longitudinal dynamics in particle accelerators. *Physics Letters, 179*, 407–413. https://doi.org/10.1016/0375-9601(93)90099-L.

Fedele, R., & Man'ko, V. I. (1999). The role of semiclassical description in the quantum-like theory of light rays. *Physical Review E, 60*, 6042–6050. https://doi.org/10.1103/PhysRevE.60.6042.

Fedele, R., Man'ko, M. A., & Man'ko, V. I. (2000). Wave-optics applications in charged-particle-beam transport. *Journal of Russian Laser Research, 21*, 1–33. https://doi.org/10.1007/BF02539473.

Fedele, R., Jovanovi, D., De Nicola, S., Mannan, A., & Tanjia, F. (2014a). Self-modulation of a relativistic charged-particle beam as thermal matter wave envelope. *Journal of Physics: Conference Series, 482*, 012014. https://doi.org/10.1088/1742-6596/482/1/012014.

Fedele, R., Tanjia, F., Jovanović, D., De Nicola, S., & Ronsivalle, C. (2014b). Wave theories of non-laminar charged particle beams: From quantum to thermal regime. *Journal of Plasma Physics, 80*, 133–145. https://doi.org/10.1017/S0022377813000913.

Feshbach, H., & Villars, F. M. H. (1958). Elementary relativistic wave mechanics of spin 0 and spin 1/2 particles. *Reviews of Modern Physics, 30*, 24–45. https://doi.org/10.1103/RevModPhys.30.24.

Fishman, L. (1992). Exact and operator rational approximate solutions of the Helmholtz, Weyl composition equation in underwater acoustics - The quadratic profile. *Journal of Mathematical Physics, 33*, 1887–1914. https://doi.org/10.1063/1.529666.

Fishman, L. (2004). One-way wave equation modeling in two-way wave propagation problems. In B. Nilsson, & L. Fishman (Eds.). *Mathematical modelling of wave phenomena 2002, mathematical modelling in physics, engineering, and cognitive sciences* (pp. 91–111). Växjö, Sweden: Växjö University Press 2004.

Fishman, L., De Hoop, M. V., & Van Stralen, M. J. N. (2000). Exact constructions of square root Helmholtz operator symbols: The focusing quadrature profile. *Journal of Mathematical Physics, 41*, 4881–4938. https://doi.org/10.1063/1.533384.

Fishman, L., & McCoy, J. J. (1984). Derivation and application of extended parabolic Wave theories. Part I. The factored Helmholtz equation. *Journal of Mathematical Physics, 25*, 285–296. https://doi.org/10.1063/1.526149.

Foldy, L. L. (1952). The electromagnetic properties of the Dirac particles. *Physical Review, 87*, 682–693. https://doi.org/10.1103/PhysRev.87.688.

Foldy, L. L., & Wouthuysen, S. A. (1950). On the Dirac theory of spin 1/2 particles and its non-relativistic Limit. *Physical Review, 78*, 29–36. https://doi.org/10.1103/PhysRev.78.29.

Forbes, G. W. (2001). Hamilton's optics: Characterizing ray mapping and opening a link to waves. *Optics and Photonics News, 12*, 34–38. https://doi.org/10.1364/OPN.12.11.000034.

Forbes, A. (2019). Structured light from lasers. *Laser and Photonics Reviews, 13*, 1900140. https://doi.org/10.1002/lpor.201900140.

Forbes, A. (2020). Structured light: Tailored for purpose. *Optics and Photonics News, 31*, 24–31. https://doi.org/10.1364/OPN.31.6.000024.

Forbes, A., de Oliveira, M., & Dennis, M. R. (2021). Structured light. *Nature Photonics, 15*, 253–262. https://doi.org/10.1038/s41566-021-00780-4.

Gerrard, A., & Burch, J. M. (1994). *Introduction to matrix methods in optics*. Dover.

Giannetto, E. (1985). A Majorana-Oppenheimer formulation of quantum electrodynamics. *Lettere al Nuovo Cimento, 44*, 140–144. https://doi.org/10.1007/BF02746912.

Gill, T. L., & Zachary, W. W. (2005). Analytic representation of the square-root operator. *Journal of Physics A: Mathematical and General, 38*, 2479–2496. https://doi.org/10.1088/0305-4470/38/11/010.

Gloge, D., & Marcuse, D. (1969). Formal quantum theory of light rays. *Journal of the Optical Society of America, 59*, 1629–1631. https://doi.org/10.1364/JOSA.59.001629.

Gomez-Reino, C., Perez, V., & Bao, C. (2002). *Gradient-index optics: Fundamentals and applications*. Springer.

Gori, F., Guattari, G., & Padovani, C. (1987). Bessel-Gauss beams. *Optics Communications, 64*, 491–495. https://doi.org/10.1016/0030-4018(87)90276-8.

Gómez-Correa, J. E., Balderas-Mata, S. E., Coello, V., Puente, N. P., Rogel-Salazar, J., & Chávez-Cerda, S. (2017). On the physics of propagating Bessel modes in cylindrical waveguides. *American Journal of Physics, 85*, 341–345. https://doi.org/10.1119/1.4976698.

Good, R. H. (1957). Particle aspect of the electromagnetic field equations. *Physical Review, 105*, 1914–1919.

Goodman, J. W. (1996). *Introduction to fourier optics* (2nd ed.). McGraw-Hill.

Gradshteyn, I. S., & Ryzhik, I. M. (2007). *Table of integrals, series, and products*. Elsevier. https://doi.org/10.1016/C2009-0-22516-5.

Greiner, W. (2000). *Relativistic quantum mechanics: Wave equations* (3rd ed.). Springer.

Greiner, W. (2001). *Quantum mechanics: An introduction* (4th ed.). Springer.

Hawkes, P. W. (2020). Dirac, c and a supper date. *Ultramicroscopy, 213*, 112981. https://doi.org/10.1016/j.ultramic.2020.112981.

Hawkes, P. W., & Kasper, E. (2017a). *Principles of electron optics - Vol. 1: Basic geometrical optics* (2nd ed.). Elsevier.

Hawkes, P. W., & Kasper, E. (2017b). *Principles of electron optics - Vol. 2: Applied geometrical optics* (2nd ed.). Elsevier.

Hawkes, P. W., & Kasper, E. (2022a). *Principles of electron optics - Vol. 3: Fundamental wave optics*. Elsevier.

Hawkes, P. W., & Kasper, E. (2022b). *Principles of electron optics - Vol. 4: Advanced wave optics*. Elsevier.

Hu, Y., Siviloglou, G., Zhang, P., Efremidis, N., Christodoulides, D. N., & Chen, Z. (2012). Self-accelerating Airy beams: Generation, control, and applications. *Nonlinear Photonics and Novel Optical Phenomena, 170*, 1–46. https://doi.org/10.1007/978-1-4614-3538-9_1.

Hui-Chuan, L., & Ji-Xiong, P. (2012). Propagation of Airy beams from right-handed material to left-handed material. *Chinese Physics B, 21*, 054201. https://doi.org/10.1088/1674-1056/21/5/054201.
Inskeep, W. H. (1988). On electromagnetic spinors and quantum theory. *Zeitschrift für Naturforschung, 43A*, 695–696. https://doi.org/10.1515/zna-1988-0715.
Ivezć, T. (2006). Lorentz invariant Majorana formulation of the field equations and Dirac-like equation for the free photon. *Electronic Journal of Theoretical Physics, 3*, 131–142. https://doi.org/10.48550/arXiv.physics/0605030.
Jackson, J. D. (1998). *Classical electrodynamics* (3rd ed.). John Wiley.
Jagannathan, R. (1990). Quantum theory of electron lenses based on the Dirac equation. *Physical Review A, 42*, 6674–6689.
Jagannathan, R. (1993). Dirac equation and electron optics. In R. Dutt, & A. K. Ray (Eds.). *Dirac and Feynman: Pioneers in Quantum Mechanics* (pp. 75–82). New Delhi, India: Wiley Eastern.
Jagannathan, R. (1999). The Dirac equation approach to spin-1/2 particle beam optics. In P. Chen (Ed.). *Proc. 15th advanced ICFA beam dynamics workshop on quantum aspects of beam physics (Monterey, California, 1998)* (pp. 670–681). World Scientific.
Jagannathan, R. (2002). Quantum mechanics of Dirac particle beam optics: Single-particle theory. In P. Chen (Ed.). *Proc. 18th advanced ICFA beam dynamics workshop on quantum aspects of beam physics (Capri, Italy, 2000)* (pp. 568–577). World Scientific.
Jagannathan, R. (2004). Quantum mechanics of Dirac particle beam transport through optical elements with straight and curved optical axes. In P. Chen, & K. Reil (Eds.). *Proc. 28th advanced ICFA beam dynamics and advanced and novel accelerators workshop (Hiroshima, Japan, 2003)* (pp. 13–21). World Scientific.
Jagannathan, R., & Khan, S. A. (1995). Wigner functions in charged particle optics. In R. Sridhar, K. SrinivasaRao, & V. Lakshminarayanan (Eds.). *Selected topics in mathematical physics – Professor R. Vasudevan memorial volume* (pp. 308–321). Delhi, India: Allied Publishers.
Jagannathan, R., & Khan, S. A. (1996). Quantum theory of the optics of charged particles. In P. W. Hawkes, B. Kazan, & T. Mulvey (Eds.). *Vol. 97. Advances in imaging and electron physics* (pp. 257–358). Academic Press.
Jagannathan, R., & Khan, S. A. (1997). Quantum mechanics of accelerator optics. *ICFA Beam Dynamics Newsletter, 13*, 21–27.
Jagannathan, R., & Khan, S. A. (2019). *Quantum mechanics of charged particle beam optics: Understanding devices from electron microscopes to particle accelerators.* Taylor & Francis.
Jagannathan, R., Simon, R., Sudarshan, E. C. G., & Mukunda, N. (1989). Quantum theory of magnetic electron lenses based on the Dirac equation. *Physical Review A, 134*, 457–464.
Jaimes-Nájera, A., Gómez-Correa, J. E., Iturbe-Castillo, M. D., Pu, J., & Chávez-Cerda, S. (2020). Kepler's law for optical beams. *Optics Express, 28*, 31979–31992. https://doi.org/10.1364/OE.403726.
Jaimes-Nájera, A., Gómez-Correa, J. E., Ugalde-Ontiveros, J. A., Méndez-Dzul, H., Iturbe-Castillo, M. S., & Chávez-Cerda, S. (2022). On the physical limitations of structured paraxial beams with orbital angular momentum. *Optics Express, 24*, 104004. https://doi.org/10.1088/2040-8986/ac84ed.
Jestädt, R., Ruggenthaler, M., Oliveira, M. J. T., Rubio, A., & Appel, H. (2019). Light-matter interactions within the Ehrenfest-Maxwell-Pauli-Kohn-Sham framework: Fundamentals, implementation, and nano-optical applications. *Advances in Physics, 68*, 225–333.
Jia, J., Fu, S., Gómez-Correa, J. E., & Chávez-Cerda, S. (2023a). *Observation of self-healing of obstructed structured beams in lenslike media. Frontiers in optics + laser science conf. proc. technical digest series.* Optica Publishing Group. https://doi.org/10.1364/FIO.2023.JTu4A.59 Paper JTu4A.59.

Jia, J., Lin, H., Fu, S., Gómez-Correa, J. E., Li, Z., Chen, Z., & Chávez-Cerda, S. (2023b). Shadows of structured beams in lenslike media. *Optics Express, 31*, 40824. https://doi.org/10.1364/OE.507030.
Khan, S. A. (1997). Quantum theory of charged particle beam optics. Ph.D. thesis (University of Madras, Chennai, India).
Khan, S. A. (1999a). Quantum theory of magnetic quadrupole lenses for spin-1/2 particles. In P. Chen (Ed.). *Proc. 15th advanced ICFA beam dynamics workshop on quantum aspects of beam physics (Monterey, California, 1998)* (pp. 670–681). World Scientific.
Khan, S. A. (1999b). Quantum aspects of accelerator optics. In A. Luccio, & W. MacKay (Eds.). *Proc. 1999 particle accelerator conference (PA99, New York)* (pp. 2817–2819).
Khan, S. A. (2002). Quantum formalism of particle beam optics. In P. Chen (Ed.). *Proc. 18th advanced ICFA beam dynamics workshop on quantum aspects of beam physics (Capri, Italy, 2000)* (pp. 568–577). World Scientific.
Khan, S. A. (2005a). An exact matrix representation of Maxwell's equations. *Physica Scripta, 71*, 440–442.
Khan, S. A. (2005b). Wavelength-dependent modifications in Helmholtz optics. *International Journal of Theoretical Physics, 44*, 95–125.
Khan, S. A. (2006). The Foldy-Wouthuysen transformation technique in optics. *Optik, 117*, 481–488.
Khan, S. A. (2007). Arab origins of the discovery of the refraction of light. *Optics and Photonics News, 18*, 22–23.
Khan, S. A. (2008). The Foldy-Wouthuysen transformation technique in optics. In P. W. Hawkes (Vol. Ed.), *Advances in imaging and electron physics. Vol. 152.* (pp. 49–78). Academic Press.
Khan, S. A. (2010). Maxwell optics of quasiparaxial beams. *Optik, 121*, 408–416.
Khan, S. A. (2014a). Aberrations in Maxwell optics. *Optik, 125*, 968–978.
Khan, S. A. (2014b). 2015 declared the international year of light and light-based technologies. *Current Science, 106*, 501. http://www.currentscience.ac.in/Volumes/106/04/0501.pdf.
Khan, S. A. (2016a). Quantum aspects of charged particle beam optics. In Al-Kamli, (Ed.). Proc. 5th Saudi International Meeting on Frontiersof Physics - 2016 (Gizan, Saudi Arabia). AIP Conference Proceedings, 1742.
Khan, S. A. (2016b). Quantum methodologies in Helmholtz optics. *Optik, 127*, 9798–9809.
Khan, S. A. (2016c). Passage from scalar to vector optics and the Mukunda-Simon-Sudarshan theory for paraxial systems. *Journal of Modern Optics, 63*, 1652–1660.
Khan, S. A. (2016d). Quantum methods in light beam optics. *Optics and Photonics News, 27*, 47.
Khan, S. A. (2016e). International year of light and history of optics. In T. Scott (Ed.). *Advances in photonics engineering, nanophotonics and biophotonics* (pp. 1–56). NOVA Science Publishers.
Khan, S. A. (2016f). Reflecting on the international year of light and light-based technologies. *Current Science, 111*, 627–631. https://doi.org/10.18520/cs/v111/i4/627-631.
Khan, S. A. (2016g). Medieval arab contributions to optics. *Digest of Middle East Studies, 25*, 19–35. https://doi.org/10.1111/dome.12065.
Khan, S. A. (2016h). Viva la fisica de México. *Physics World, 29*, 21–22. https://doi.org/10.1088/2058-7058/29/1/27.
Khan, S. A. (2017a). Linearization of wave equations. *Optik, 131*, 350–363.
Khan, S. A. (2017b). Hamilton's optical-mechanical analogy in the wavelength-dependent regime. *Optik, 130*, 714–722.
Khan, S. A. (2017c). Polarization in Maxwell optics. *Optik, 131*, 733–748.

Khan, S. A. (2017d). Medieval arab achievements in optics. In R. Roshdi, B. Azzedine, & V. Lakshminarayanan (Eds.). *Light-based science: Technology and sustainable development. Proc. Islamic golden age of science for today's knowledge-based society: The Ibn Al-Haytham Example (Paris, 2015).* Taylor & Francis. http://isbn.nu/9781498779388.

Khan, S. A. (2017e). Quantum methodologies in Maxwell optics. In P. W. Hawkes (Vol. Ed.), *Advances in imaging and electron physics. Vol. 201.* (pp. 57–135). Academic Press.

Khan, S. A. (2018a). Aberrations in Helmholtz optics. *Optik, 153*, 164–181.

Khan, S. A. (2018b). Quantum mechanical techniques in light optics. In Al-Kamli, (Ed.). *Proc. 6th International Meeting on Frontiers of Physics - 2018 (Gizan, Saudi Arabia, 2018) AIP Conf, Proc. 1976* (pp. 020016).

Khan, S. A. (2018c). E.C.G. Sudarshan and the quantum mechanics of charged-particle beam optics. *Current Science, 115*, 1813–1814.

Khan, S. A. (2020). Quantum mechanical techniques in light beam optics. *Frontiers in Optics JTu1B*, 39.

Khan, S. A. (2021). Cross polarization in Gaussian light beams. Optics InfoBase Conference Papers, art. no. JW7A.53.

Khan, S. A. (2022). Cross polarization in bessel light beams. Optics InfoBase Conference Papers, art. no. JW4B.65.,

Khan, S. A. (2023a). Cross polarization in gaussian and bessel light beams. *Optics Communications, 545*, 129728.

Khan, S. A. (2023b). A matrix differential operator for passage from scalar to vector wave 0ptics. *Results in Optics, 13*, 100527.

Khan, S. A. (2023c). Anisotropic Airy beams. *Results in Optics, 13*, 100569.

Khan, S. A. (2023). Cross polarization in Bessel-Gaussian light beams. Optics InfoBase Conference Papers, art. no. JM7A.59.

Khan, S. A. (2024). Cross polarization in anisotropic Gaussian light beams. *Indian Journal of Physics, 98*, 3699–3705.

Khan, S. A., & Jagannathan, R. (1995). Quantum mechanics of charged particle beam transport through magnetic lenses. *Physical Review E, 51*, 2510–2515.

Khan, S. A., & Jagannathan, R. (2020). Quantum mechanics of bending of a nonrelativistic charged particle beam by a dipole magnet. *Optik, 206*, 163626.

Khan, S. A., & Jagannathan, R. (2021). Quantum mechanics of round magnetic electron lenses with Glaser and power law models of *B(z)*. *Optik, 229*, 166303.

Khan, S. A., & Jagannathan, R. (2024a). Classical and quantum mechanics of the Wien velocity filter. *International Journal of Theoretical Physics, 63*, 2024.

Khan, S. A., & Jagannathan, R. (2024b). In M. Hÿtch, & P. W. Hawkes (Eds.). *Quantum mechanics of bending of a charged particle beam by a dipole magnet. Vol. 229.* (pp. 1–41) Elsevier.

Khan, S. A., & Jagannathan, R. (2024c). A new matrix representation of the Maxwell equations based on the Riemann-Silberstein-Weber vector for a linear inhomogeneous medium. *Results in Optics, 17*, 100747.

Khan, S. A., Jagannathan, R., & Simon, R. (2002). Foldy-Wouthuysen transformation and a quasiparaxial approximation scheme for the scalar wave theory of light beams. arXiv:physics/0209082 [physics.optics].

Khan, S. A., & Pusterla, M. (1999). Quantum mechanical aspects of the halo puzzle. In A. Luccio, & W. MacKay (Eds.). Proc. 1999 Particle Accelerator Conference (PA99, New York).

Khan, S. A., & Pusterla, M. (2000a). Quantum-like approaches to the beam halo problem. In D. Han, D. Kim, & Y. S. Solimeno (Eds.). http://arxivorg/abs/physics/9905034 Proc. 6th International Conf. on Squeezed States and Uncertainty Relations, Napoli, Italy, 1999. NASA Conference Publication Series, 2000-209899. arXiv:physics/9905034 [physics.acc-ph].

Khan, S. A., & Pusterla, M. (2000b). Quantum-like approach to the transversal and longitudinal beam dynamics. The halo problem. *European Physical Journal A, 7*, 583–587. https://doi.org/10.1007/s100500050430.

Khan, S. A., & Pusterla, M. (2001). Quantum approach to the halo formation in high current beams. *Nuclear Instruments and Methods in Physics Research, 464*, 461–464. https://doi.org/10.1016/S0168-9002(01)00108-5.

Khan, S. A., & Wolf, K. B. (2002). Hamiltonian orbit structure of the set of paraxial optical systems. *Journal of the Optical Society of America, 19*, 2436–2444. https://doi.org/10.1364/JOSAA.19.002436.

Kiesslinga, MK-H., & Tahvildar-Zadehb, A. S. (2018). On the quantum mechanics of a single photon. *Journal of Mathematical Physics, 59*, 112302.

Kisel, V. V., Ovsiyuk, E. M., Red'kov, V. M., & Tokarevskaya, N. G. (2011). Maxwell equations in complex form, squaring procedure and separating the variables. *Ricerche di Matematica, 60*, 1–14.

Kogelnik, H. (1965). On the propagation of Gaussian beams of light through lenslike media including those with a loss or gain variation. *Applied Optics, 4*, 1562–1569. https://doi.org/10.1364/AO.4.001562.

Kogelink, H., & Li, T. (1966). Laser beams and resonators. *Applied Optics, 5*, 1550–1567. https://doi.org/10.1364/AO.5.001550.

Korotkova, O., & Testorf, M. (2023). Introducing JOSA A retrospectives: Editorial. *Journal of the Optical Society of America, 40*, ED3–ED4.

Kulyabov, D. S. (2016). Spinor-like Hamiltonian for Maxwellian optics. In Gh. Adam, J. Buša, & M. Hnatič (Vol. Eds.), *Proc. mathematical modelingand computational physics (stará Lesná, Slovakia, 2015). EPJ web of conferences, Vol. 108,* id.02034. https://doi.org/10.1051/epjconf/201610802034.

Kulyabov, D. S., Korolkova, A. V., & Sevastianov, L. A. (2017). Spinor representation of Maxwell's equations. *Journal of Physics: Conference Series, 788*, 012025.

Lakshminarayanan, V., & Varadharajan, L. S. (2015). *Special functions for optical science and engineering*. SPIE Press. https://doi.org/10.1117/3.2207310.

Lakshminarayanan, L., Ghatak, A., & Thyagarajan, K. (2002). *Lagrangian optics*. Springer. https://doi.org/10.1007/978-1-4615-1711-5.

Lakshminarayanan, V., Sridhar, R., & Jagannathan, R. (1998). Lie algebraic treatment of dioptric power and optical aberrations. *Journal of the Optical Society of America, 15*, 2497–2503. https://doi.org/10.1364/JOSAA.15.002497.

Laporte, O., & Uhlenbeck, G. E. (1931). Applications of spinor analysis to the Maxwell and Dirac Equations. *Physical Review, 37*, 1380–1397. https://doi.org/10.1103/PhysRev.37.1380.

Lax, M., Louisell, W. H., & McKnight, W. B. (1975). From Maxwell to paraxial wave optics. *Physical Review A, 11*, 1365–1370. https://doi.org/10.1103/PhysRevA.11.1365.

Livadiotis, G. (2018). Complex symmetric formulation of Maxwell equations for fields and potentials. *Mathematics, 6*, 114.

Li, Y., Lee, H., & Wolf, E. (2004). New generalized Bessel-Gaussian beams. *Journal of the Optical Society of America, 21*, 640–646. https://doi.org/10.1364/JOSAA.21.000640.

Magnus, W. (1954). On the exponential solution of differential equations for a linear operator. *Communications on Pure and Applied Mathematics, 7*, 649–673.

Mananga, E. S., & Charpentier, T. (2016). On the Floquet-Magnus expansion: Applications in solid-state nuclear magnetic resonance and physics. *Physics Reports, 609*, 1–49. https://doi.org/10.1016/j.physrep.2015.10.005.

Mandel, L., & Wolf, E. (1995). *Optical coherence and quantum optics*. Cambridge University Presshttps://doi.org/10.1017/CBO9781139644105.

Mazharimousavi, S. H., Roozbeh, A., & Halilsoy, M. (2013). Electromagnetic wave propagation through inhomogeneous material layers. *Journal of Electromagnetic Waves and Applications, 27*, 2065–2074. https://doi.org/10.1080/09205071.2013.831741.

Mehrafarin, M., & Balajany, H. (2010). Paraxial spin transport using the Dirac-like paraxial wave equation. *Physics Letters, 374*, 1608–1610. https://doi.org/10.1016/j.physleta.2010.01.067.
Mignani, R., Recami, E., & Baldo, M. (1974). About a Dirac-like equation for the photon, according to Ettore Majorana. *Lettere al Nuovo Cimento, 11*, 568–572. https://doi.org/10.1007/BF02812391.
Mingjie, L., Peng, S., Luping, D., & Xiaocong, Y. (2020). Electronic Maxwell's equations. *New Journal of Physics, 22*, 113019.
Mishra, S. R. (1991). A vector wave analysis of a Bessel beam. *Optics Communications, 85*, 159–161. https://doi.org/10.1016/0030-4018(91)90386-R.
Mohr, P. J. (2010). Solutions of the Maxwell equations and photon wave functions. *Annals of Physics, 325*, 607–663. https://doi.org/10.1016/j.aop.2009.11.007.
Mondragón, S. J., & Wolf, K. B. (1986). Lie methods in optics, *Lecture Notes in Physics, 250*, https://doi.org/10.1007/3-540-16471-5.
Moses, E. (1959). Solutions of Maxwell's equations in terms of a spinor notation: The direct and inverse problems. *Physical Review, 113*, 1670–1679. https://doi.org/10.1103/PhysRev.113.1670.
Mukunda, N., Simon, R., & Sudarshan, E. C. G. (1983). Paraxial-wave optics and relativistic front description II: The vector theory. *Physical Review A, 28*, 2933–2942. https://doi.org/10.1103/PhysRevA.28.2933.
Mukunda, N., Simon, R., & Sudarshan, E. C. G. (1985a). Fourier optics for the Maxwell field: Formalism and applications. *Journal of the Optical Society of America, 2*, 416–426. https://doi.org/10.1364/JOSAA.2.000416.
Mukunda, N., Simon, R., & Sudarshan, E. C. G. (1985b). Paraxial Maxwell beams: Transformation by general linear optical systems. *Journal of the Optical Society of America, 2*, 1291–1296. https://doi.org/10.1364/JOSAA.2.001291.
Nazarathy, M., & Shamir, J. (1980). Fourier optics described by operator algebra. *Journal of the Optical Society of America A, 70*, 150–159. https://doi.org/10.1364/JOSA.70.000150.
Nazarathy, M., & Shamir, J. (1982). First-order optics - a canonical operator representation: lossless systems. *Journal of the Optical Society of America A, 72*, 356–364. https://doi.org/10.1364/JOSA.72.000356.
Nomoto, S., Aadhi, A., Prabhakar, S., Singh, R. P., Vyas, R. P., & Singh,S, R. (2015). Polarization properties of the Airy beam. *Optics Letters, 40*, 4516–4519. https://doi.org/10.1364/OL.40.004516.
Ohmura, T. (1956). A new formulation on the electromagnetic field. *Progress of Theoretical and Experimental Physics, 16*, 684–685. https://doi.org/10.1143/PTP.16.684.
Oppenheimer, J. R. (1931). Note on light quanta and the electromagnetic field. *Physical Review, 38*, 725–746. https://doi.org/10.1103/PhysRev.38.725.
Orris, G. J., & Wurmser, D. (1995). Applications of the Foldy-Wouthuysen transformation to acoustic modeling using the parabolic equation method. *Journal of the Acoustical Society of America, 98*, 2870. https://doi.org/10.1121/1.413215.
Osche, G. R. (1977). Dirac and Dirac-Pauli equation in the Foldy-Wouthuysen representation. *Physical Review D, 15*, 2181–2185. https://doi.org/10.1103/PhysRevD.15.2181.
Otte, E. (2020). *Structured singular light fields*. Springer. https://doi.org/10.1007/978-3-030-63715-6.
Ovsiyuk, E. M., Kisel, V. V., & Red'kov, V. M. (2013). *Maxwell electrodynamics and Boson fields in spaces of constant curvature*. NOVA Science Publishers. http://www.novapublishers.com/.
Panofsky, W. K. H., & Phillips, M. (1962). *Classical electricity and magnetics*. Addison-Wesley.

Petroni, N. C., De Martino, S., De Siena, S., & Illuminati, F. (2000). Stochastic collective dynamics of charged-particle beams in the stability regime. *Physical Review E, 63*, 016501. https://doi.org/10.1103/PhysRevE.63.016501.

Pradhan, T. (1987). Maxwell's equations from geometrical optics. *Physics Letters A, 122*, 397–398. https://doi.org/10.1016/0375-9601(87)90735-3.

Ram, A. K., Vahala, G., Vahala, L., & Soe, M. (2021). Reflection and transmission of electromagnetic pulses at a planar dielectric interface: Theory and quantum lattice simulations. *AIP Advances, 11*, 105116.

Rangarajan, G., & Sachidanand, M. (1997). Spherical aberrations and its correction using Lie algebraic methods. *Pramana, 49*, 635–643. https://doi.org/10.1007/BF02848337.

Rangarajan, G., & Sridharan, S. (2010). Invariant norm quantifying nonlinear content of Hamiltonian systems. *Applied Mathematics and Computation, 217*, 2495–2500. https://doi.org/10.1016/j.amc.2010.07.060.

Rangarajan, G., Dragt, A. J., & Neri, F. (1990). Solvable map representation of a nonlinear symplectic map. *Part. Accel, 28*, 119–124. http://cds.cern.ch/record/1108134/files/p119.pdf.

Red'kov, V. M., Tokarevskaya, N. G., & Spix, G. J. (2012). Majorana-Oppenheimer approach to Maxwell electrodynamics. Part I. Minkowski space. *Advances in Applied Clifford Algebras, 22*, 1129–1149. https://doi.org/10.1007/s00006-012-0320-1.

Robson, C. W., Tamashevich, Y., Rantala, T. T., & Ornigotti, M. (2021). Path integrals: From quantum mechanics to photonics. *APL Photonics, 6*, 071103. https://doi.org/10.1063/5.0055815.

Rogel-Salazar, J., Jiménez-Romero, H. A., & Chávez-Cerda, S. (2014). Full characterization of Airy beams under physical principles. *Physical Review A, 89*, 023807. https://doi.org/10.1103/PhysRevA.89.023807 Erratum Phys. Rev. A 97 (2018) 059901.

Rubinsztein-Dunlop, H., Forbes, A., Berry, M. V., Dennis, M. R., Andrews, D. L., Mansuripur, M., ... Bauer, T. (2017). Roadmap on structured light. *Journal of Optics, 19*, 013001. https://doi.org/10.1088/2040-8978/19/1/013001 .

Sachs, M., & Schwebel, S. L. (1962). On covariant formulations of the Maxwell-Lorentz theory of electromagnetism. *Journal of Mathematical Physics, 3*, 843–848. https://doi.org/10.1063/1.1724297 .

Saghafi, S., Sheppard, C. J. R., & Piper, J. A. (2001). Characterising elegant and standard Hermite-Gaussian beam modes. *Optics Communications, 191*, 173–179. https://doi.org/10.1016/S0030-4018(01)01110-5.

Sebens, C. T. (2019). Electromagnetism as quantum physics. *Foundations of Physics, 49*, 365–389.

Seshadri, S. R. (2009). Linearly polarized anisotropic Gaussian light wave. *Journal of the Optical Society of America A, A26*, 1582.

Sheppard, C. J. R., Rehman, S., Balla, N. K., Yew, E. Y. S., & Teng, T. W. (2009). Bessel beams: Effects of polarization. *Optics Communications, 282*, 4647–4656. https://doi.org/10.1016/j.optcom.2009.08.058.

Shou, X., Bai, L., & Guo, X. (2011). Single-lens equivalent systems for Grin lenses. *Optik, 122*, 827–835.

Silberstein, L. (1907a). Elektromagnetische Grundgleichungen in bivektorieller Behandlung. *Annals of Physics (Leipzig), 327*, 579–586. https://doi.org/10.1002/andp.19073270313.

Silberstein, L. (1907b). Nachtrag zur Abhandlung über Elektromagnetische Grundgleichungen in bivektorieller Behandlung. *Annals of Physics, 329*, 783–784. https://doi.org/10.1002/andp.19073291409.

Silenko, A. J. (2016). Exact form of the exponential Foldy-Wouthuysen transformation operator for an arbitrary-spin particle. *Physical Review A, 94*, 032104. https://doi.org/10.1103/PhysRevA.94.032104.

Simon, R. (1983). Anisotropic Gaussian beams. *Optics Communications, 46*, 265–269.

Simon, R. (1985). A new class of anisotropic Gaussian beams. *Optics Communications, 55*, 381–385.
Simon, R. (1987). Laser cavities bounded by crossed cylindrical mirrors. *Journal of the Optical Society of America, 4*, 1953.
Simon, R., & Mukunda, N. (1998). Iwasawa decomposition in first-order optics: Universal treatment of shape-invariant propagation for coherent and partially coherent beams. *Journal of the Optical Society of America, 15*, 2146–2155. https://doi.org/10.1364/JOSAA.15.002146.
Simon, R., Sudarshan, E. C. G., & Mukunda, N. (1986). Gaussian-Maxwell beams. *Journal of the Optical Society of America, 4*, 536–540. https://doi.org/10.1364/JOSAA.3.000536.
Simon, R., Sudarshan, E. C. G., & Mukunda, N. (1987). Cross polarization in laser beams. *Applied Optics, 26*, 1589–1593. https://doi.org/10.1364/AO.26.001589.
Siviloglou, G. A., Broky, J., Dogariu, A., & Christodoulides, D. N. (2007). Observation of accelerating Airy beams. *Physical Review Letters, 99*(2007), 213901. https://doi.org/10.1103/PhysRevLett.99.213901.
Siviloglou, G. A., Broky, J., Dogariu, A., & Christodoulides, D. N. (2008). Ballistic dynamics of Airy beams. *Optics Letters, 33*, 207–209. https://doi.org/10.1364/OL.33.000207.
Spiegel, M. R., & Liu, J. (1999). *Mathematical handbook of formulas and tables: Schaum's outlines*. McGraw-Hill. http://isbn.nu/9780071777476/.
Sudarshan, E. C. G., Simon, R., & Mukunda, N. (1983). Paraxial-wave optics and relativistic front description I: The scalar theory. *Physical Review A, 28*, 2921–2932. https://doi.org/10.1103/PhysRevA.28.2921.
Tsai, K., & Chu, S. (2013). A method to calculate arbitrary linear polarized laser beam evolutions in GRIN lenses. *Proceedings of SPIE, 8637*, 86370N–1.
Ugalde-Ontiveros, J. A., Jaimes-Nájera, A., Luo, S., Gómez-Correa, J. E., Pu, J., & Chávez-Cerda, S. (2021a). What are the traveling waves composing the Hermite-Gauss beams that make them structured wavefields? *Optics Express, 29*, 29068–29081. https://doi.org/10.1364/OE.424782.
Ugalde-Ontiveros, J. A., Jaimes-Nájera, A., Gómez-Correa, J. E., & Chávez-Cerda, S. (2021b). Siegman's elegant laser resonator modes. *Optics and Laser Technology, 143*, 107340. https://doi.org/10.1016/j.optlastec.2021.107340.
Vahala, G., Vahala, L., Soe, M., & Ram, A. K. (2020a). Unitary quantum lattice simulations for Maxwell equations in vacuum and in dielectric media. *Journal of Plasma Physics, 86*, 905860518. https://doi.org/10.1017/S0022377820001166.
Vahala, G., Vahala, L., Soe, M., & Ram, A. K. (2020b). Building a three-dimensional quantum lattice algorithm for Maxwell equations. *Radiation Effects and Defects in Solids, 175*, 986–990.
Vahala, G., Vahala, L., Soe, M., & Ram, A. K. (2020). The effect of the Pauli spin matrices on the quantum lattice algorithm for Maxwell equations in inhomogeneous media. arXiv:2010.12264[physics.plasm-ph]. https://arxiv.org/abs/2010.12264
Vahala, G., Vahala, L., Soe, M., & Ram, A. K. (2021a). One and two-dimensional quantum lattice algorithms for Maxwell equations in inhomogeneous scalar dielectric media I: Theory. *Radiation Effects and Defects in Solids, 176*, 49–63.
Vahala, G., Vahala, L., Soe, M., & Ram, A. K. (2021b). One and two-dimensional quantum lattice algorithms for Maxwell equations in inhomogeneous scalar dielectric media II: Simulations. *Radiation Effects and Defects in Solids, 176*, 64–72.
Vahala, G., Vahala, L., Ram, A. K., & Soe, M. (2022). The effect of the width of the incident pulse to the dielectric transition layer in the scattering of an electromagnetic pulse. arXiv:2201.09259[physics.plasm-ph]. https://arxiv.org/abs/2201.09259.

Vahala, G., Hawthorne, J., Vahala, L., Ram, A. K., & Soe, M. (2022). Quantum lattice representation for the curl equations of Maxwell equations. *Radiation Effects and Defects in Solids, 177*, 85–94.

Wang, Z.-Y. (2015). The (1, 0) ⊕ (0, 1) spinor description of the photon field and its applications. arXiv:1508.02321[quant-phys]. http://arxiv.org/abs/1508.02321.

Wilcox, R. M. (1967). Exponential operators and parameter differentiation in quantum physics. *Journal of Mathematical Physics, 8*, 962–982. https://doi.org/10.1063/1.1705306.

Wolf, K. B. (1988). Diffraction-free beams remain diffraction free under all paraxial optical transformations. *Physical Review Letters, 60*, 757–759. https://doi.org/10.1103/PhysRevLett.60.757.

Wolf, K. B. (1989). Lie methods in optics. *Lecture Notes in Physics,* 352. https://doi.org/10.1007/BFb0012741.

Wünsche, A. (1992). Transition from the paraxial approximation to exact solutions of the wave equation and application to Gaussian beams. *Journal of the Optical Society of America, 9*, 765–774. https://doi.org/10.1364/JOSAA.9.000765.

Zhang, P., Prakash, J., Zhang, Z., Mills, M., Efremidis, N., Christodoulides, D., & Chen, Z. (2011). Trapping and guiding microparticles with morphing autofocusing Airy beams. *Optics Letters, 36*, 2883–2885. https://doi.org/10.1364/OL.36.002883.

CHAPTER TWO

Generalized quantum theory of non-paraxial interference

Román Castañeda*
Physics Department, Universidad Nacional de Colombia Sede Medellín, Medellín, Colombia
*Corresponding author. e-mail address: rcastane@unal.edu.co

Contents

Abstract

By introducing the quantum interference operator, the phenomenology of the non-paraxial interference in ordinary space is discussed. It is shown that the same principle of confinement in geometric states of space accounts for single photon and single matter particle interference, and that such geometric space states live in vacuum and can characterize completely the preparation-and-measurement configured interferometers. Although it is a corpuscular framework, interference with classical light is also approached accurately. Therefore, this new theory is exact (non-paraxial), generalized and exhaustive. Its phenomenology requires fewer basic premises than the standard formalisms and

Advances in Imaging and Electron Physics, Volume 235
ISSN 1076-5670, https://doi.org/10.1016/bs.aiep.2025.06.001

restores the ordinary space as the environment in which the individual experimental realizations of quantum interference take place, thus solving the Feynman's only mystery of quantum mechanics. Further phenomenological implications of the quantum interference operator are discussed too.

1. Introduction

Interference is a basic phenomenon of physics at the foundations of our understanding of nature. Conventionally, this phenomenon is attributed to the wave nature of light and matter, and practically the same mathematical apparatus, based on the wave superposition, is used to formalize interference with classical light, single photons and single matter particles (Born & Wolf, 1993; Feynman et al., 1965; Mandel & Wolf, 1995). Nevertheless, phenomenological descriptions of interference observed with the most basic interferometers are not unified. Young introduced the wave superposition as the phenomenological principle for light interference in ordinary space (Young, 1804), while the superposition of quantum wave functions in Hilbert space is considered an accurate calculation procedure, without a precise physical meaning, for predicting experimental interference outcomes with single photons and single matter particles (De Martini et al., 1994). In this context, wave—particle duality and related hypotheses, like the particle delocalization, the self-interference and the quantum wave function collapse, have been introduced to support phenomenological approaches of these interference cases (De Martini et al., 1994; Tang & Hu, 2022). In addition, the mathematical model is usually paraxial approximated, so that its accuracy is restricted to the far-field (Born & Wolf, 1993).

Therefore, the formulation of an exact (non-paraxial) generalized theory of interference in ordinary space, based on a unique phenomenological principle and without resorting to additional premises or hypothesis, seems to be still an open question in the context of physics foundations (Dong-Yeop & Weng Cho, 2020; Eberly et al., 2016; Wódkiewicz & Herling, 1998). Such a theory should explain single photon and single matter particle interference in the same corpuscular framework in ordinary space, and should provide an accurate approach to classical light interference. In addition, the theory should be focused on the measurement by square modulus detectors.

To propose a theory that meets such features is the main subject of the current paper. It is developed both mathematically and phenomenologically. The mathematical fundamentals are deduced in detail in Section 2, with support on the conventional quantization of Maxwell's equations in free space (Scully & Zubairy, 1997; Shih, 2020) and the Schrödinger equation for field-free regions (Feynman et al., 1965).

The main mathematical object of the theory is the quantum interference operator, discussed in Section 3. It is deduced in space-time, and realizes the principle of confinement in geometric states of space (Castañeda et al., 2023), thus involving only the density operator of geometric states of space and the number operator of single photons or matter particles confined in the space states. Peculiar phenomenological implications of the quantum interference operator are discussed too, such as the role of vacuum in interference and the description of the individual experimental realization in the conventional P&M (preparation and measurement) interferometers. Any individual experimental realization is the segment that begins with the single particle or photon emission and ends with its detection, so that only one particle or photon moves in the interferometer, without connection or influence of particles or photons in previous individual realizations.

In spite of its corpuscular meaning, the theory provides also an interesting approach to classical light interference, discussed in Section 4. Theory connections with reported features, such as spatial entanglement, diffraction and geometric uncertainty (Castañeda et al., 2023), are also discussed in Section 5.

In this way, the phenomenological generalization and exhaustivity in ordinary space of the proposed quantum theory of non-paraxial interference is demonstrated, as well as its mathematical accuracy. Because of these features, such a theory is a powerful tool that improves the understanding of interference fundamentals and its experimental implementation.

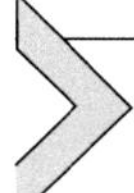

2. Non-paraxial propagation in ordinary space

2.1 Single photons

Let us determine the quantum Hamiltonian of single photons that non-paraxially propagate in ordinary space. To this aim, the start point is the Hamiltonian of the electromagnetic field $H = \frac{1}{2}\int_V d^3r \left(\varepsilon_0 \,|\mathbf{E}(\mathbf{r},\, t)|^2 + \frac{1}{\mu_0}\,|\mathbf{B}(\mathbf{r},\, t)|^2\right)$, where the electric and magnetic fields can be expressed as $\mathbf{E}(\mathbf{r},\, t) = -\frac{\partial}{\partial t}\mathbf{A}(\mathbf{r},\, t)$ and $\mathbf{B}(\mathbf{r},\, t) = \nabla \times \mathbf{A}(\mathbf{r},\, t)$, respectively. $\mathbf{A}(\mathbf{r}, t)$ is the vector potential that fulfils Coulomb gauge. Consequently, $\nabla^2\mathbf{A}(\mathbf{r}, t) - \frac{1}{c^2}\frac{\partial^2\mathbf{A}(\mathbf{r}, t)}{\partial t^2} = 0$, with c the light speed in vacuum, whose solution takes the general form

$$\mathbf{A}(\mathbf{r}, t) = \sum_n \mathbf{A}_n(\mathbf{r})\left(\hat{a}_n^\dagger \exp(i\omega_n t) + \hat{a}_n \exp(-i\omega_n t)\right) \tag{1}$$

in quantum notation (Scully & Zubairy, 1997; Shih, 2020; Walls & Milburn, 1995). In Eq. (1), ω_n denotes the frequencies of the photon spectral band and $\hat{a}_n^\dagger$, $\hat{a}_n$ represent the creation and annihilation operators in each spectral mode, which fulfil the bosonic commutation rules $[\hat{a}_n, \hat{a}_n] = [\hat{a}_n^\dagger, \hat{a}_n^\dagger] = 0$ and $[\hat{a}_n, \hat{a}_n^\dagger] = 1$ (Walls & Milburn, 1995). The spatial coefficient is the solution to Helmholtz equation $\nabla^2\mathbf{A}_n(\mathbf{r}) + k_n^2\mathbf{A}_n(\mathbf{r}) = 0$, with $k_n = \omega_n/c$.

In the standard quantum optics formalism (Scully & Zubairy, 1997; Shih, 2020; Walls & Milburn, 1995), $\mathbf{A}_n(\mathbf{r}) = C_n\ \mathbf{e}_n \exp(i\ \mathbf{k}_n\cdot\mathbf{r})$, with $C_n = \left(\frac{\hbar}{2\omega_n\varepsilon_0}\right)^{\frac{1}{2}}$, $\hbar = h/2\pi$ (h Planck constant) and ε_0 the electric permittivity of vacuum. The unit polarization vector $\mathbf{e}_n$ of each mode is orthogonal to the propagation vector $\mathbf{k}_n$ whose magnitude is discretized in a cubic region of space of side L, corresponding to the physical volume delimited by planes M and D (Fig. 1), in accordance to the rule for standing waves $|\mathbf{k}_n| = k_n = \frac{2\pi n}{L}$, with $n = \sqrt{n_x^2 + n_y^2 + n_z^2}$, $n_j = 0, \pm1, \pm2, \cdots$ and $j = x, y, z$ (Walls & Milburn, 1995). So, the Hamiltonian becomes

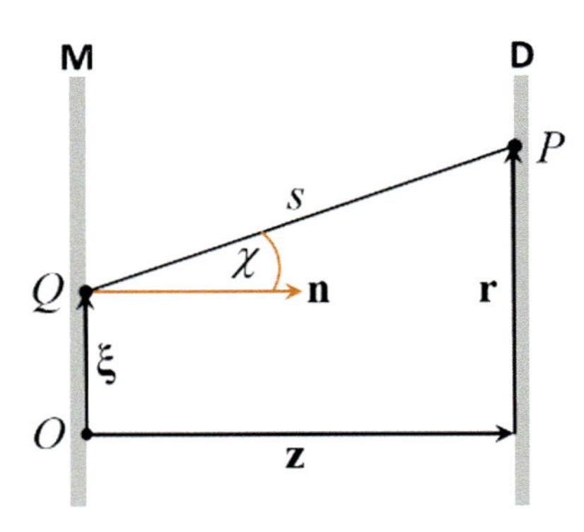

Fig. 1 Conceptual sketch of single-photon propagation in ordinary space.

$H = \sum_n \hbar\omega_n \left(\hat{a}_n^\dagger \hat{a}_n + \frac{1}{2}\right)$ (Walls & Milburn, 1995), with $\hat{a}_n^\dagger \hat{a}_n = \hat{n}_n$ the number operator that specifies the number of photons $n_n = \langle n_n | \hat{n}_n | n_n \rangle$ with the frequency ω_n. The term $\frac{1}{2}\hbar\omega_n$ is interpreted as the energy of the vacuum fluctuation in each mode (Shih, 2020; Walls & Milburn, 1995).

However, the eigenfunctions $\exp(i\ \mathbf{k}_n \cdot \mathbf{r})$ characterize the approximated solution called the paraxial approach (Born & Wolf, 1993), whose phenomenological validity is restricted to the far-field propagation. In order to describe the photon behavior in the complete volume for any distance between M and D, the exact solution of Helmholtz equation should be regarded. It is obtained by applying Green's theorem and Kirchhoff's boundary conditions to Helmholtz equation (Born & Wolf, 1993; Castañeda et al., 2020), thus allowing expressing the exact solution as the expansion

$$\mathbf{A}_n(\mathbf{r}) = \int_M d^2\xi\ \mathbf{A}_n(\boldsymbol{\xi})\ \Theta_n(\boldsymbol{\xi}, \mathbf{r}, \mathbf{z}, k_n) \tag{2}$$

on the non-paraxial basis of eigenfunctions of the Laplacian operator (Castañeda et al., 2020)

$$\Theta_n(\boldsymbol{\xi}, \mathbf{r}, \mathbf{z}, k_n) = -i\frac{k_n}{4\pi}\left(\frac{z + |\mathbf{z} + \mathbf{r} - \boldsymbol{\xi}|}{|\mathbf{z} + \mathbf{r} - \boldsymbol{\xi}|^2}\right) \exp(ik_n |\mathbf{z} + \mathbf{r} - \boldsymbol{\xi}|). \tag{3}$$

Eq. (2) is compatible with the discretization of the propagation vector and its expansion coefficient takes the form $\mathbf{A}_n(\boldsymbol{\xi}) = \mathbf{A}_n^0(\boldsymbol{\xi})\ t(\boldsymbol{\xi})$, whose factors are boundary conditions on M. Specifically, $t(\boldsymbol{\xi}) = |t(\boldsymbol{\xi})| \exp(i\phi(\boldsymbol{\xi}))$ denotes the complex transmission function at each point of M, that implicitly includes all quantum interactions between the photon and the interference device, usually a mask, and $\mathbf{A}_n^{(0)}(\boldsymbol{\xi}) = \left(\frac{\hbar}{2\omega_n \varepsilon_0}\right)^{\frac{1}{2}} \psi_n(\boldsymbol{\xi})\ \mathbf{e}_n(\boldsymbol{\xi})$ represents the prepared vector potential at M, with $\psi_n(\boldsymbol{\xi})$ an eigenfunction of the Laplacian in Helmholtz equation that determines the complex amplitude of probability of photons of frequency ω_n at each point of M. Therefore, Eq. (2) becomes

$$\mathbf{A}_n(\mathbf{r}) = \left(\frac{\hbar}{2\omega_n \varepsilon_0}\right)^{\frac{1}{2}} \int_M d^2\xi\ \psi_n(\boldsymbol{\xi})\ t(\boldsymbol{\xi})\ \mathbf{e}_n(\boldsymbol{\xi})\ \Theta_n(\boldsymbol{\xi}, \mathbf{r}, \mathbf{z}, k_n). \tag{4}$$

By regarding the local connection between electric and magnetic field vectors of light, the Hamiltonian can be expressed as $H = \varepsilon_0 \int_V d^3r \,|\mathbf{E}(\mathbf{r}, t)|^2$ (Born & Wolf, 1993), with $\mathbf{E}(\mathbf{r}, t) = \sum_n \mathbf{E}_n(\mathbf{r}, t)$, i.e.

$$\mathbf{E}(\mathbf{r}, t) = -i \sum_n \left(\frac{\hbar\omega_n}{2\varepsilon_0}\right)^{\frac{1}{2}} \left(\hat{a}_n^\dagger \exp(i\omega_n t) - \hat{a}_n \exp(-i\omega_n t)\right) \int_M d^2\xi \;\psi_n(\boldsymbol{\xi})\; t(\boldsymbol{\xi})\, \mathbf{e}_n(\boldsymbol{\xi})\; \Theta_n(\boldsymbol{\xi}, \mathbf{r}, \mathbf{z}, k_n), \tag{5}$$

and therefore $|\mathbf{E}(\mathbf{r}, t)|^2 = \sum_n \sum_m \mathbf{E}_n(\mathbf{r}, t)\cdot\mathbf{E}_m^*(\mathbf{r}, t)$.

Nevertheless, in each individual experimental realization only one photon of energy $E = \hbar\omega$ moves in the setup. Thus, it should be considered as a monochromatic event even though the specific energy value is limited by the uncertainty principle $\Delta E \;\Delta t \geq \hbar/2$. Furthermore, single photons of different experimental realizations are not connected in any way. Therefore, $|\mathbf{E}(\mathbf{r}, t)|^2 = \sum_n |\mathbf{E}_n(\mathbf{r}, t)|^2$ if polychromatic sources of single photons are used. It reduces to only the term for $\omega_n = \omega$ by quasi-monochromatic sources of main frequency ω.

In addition, the two terms of the time component of $\mathbf{E}_n(\mathbf{r}, t)$ denote two degrees of freedom in orientation, corresponding to opposite directions of propagation. However, because of photon annihilation at D (Fig. 1), such degrees of freedom represent the two possible orientations of the setup optical axis, so that the corresponding directions of propagation are independent to each other. This means that the two terms of the time component do not overlap, and therefore the cross-terms of their product in $|\mathbf{E}_n(\mathbf{r}, t)|^2$ nullify.

It is also well known that single photon detection occurs at a given pixel of the square modulus detector placed at D. This means that single photon measurement is essentially local, i.e. the energy to be measured is contained in a volume of size arbitrarily smaller than the pixel size, in such a way that the Hamiltonian becomes $H = \varepsilon_0 |\mathbf{E}(\mathbf{r}, t)|^2 = \varepsilon_0 \sum_n |\mathbf{E}_n(\mathbf{r}, t)|^2$. It is important to clarify that technical parameters involved in the detector-photon interaction, like detector responsivity and integration time, are not addressed in this paper because the paper main subject (i.e. the interference phenomenology) is independent of detector specifications.

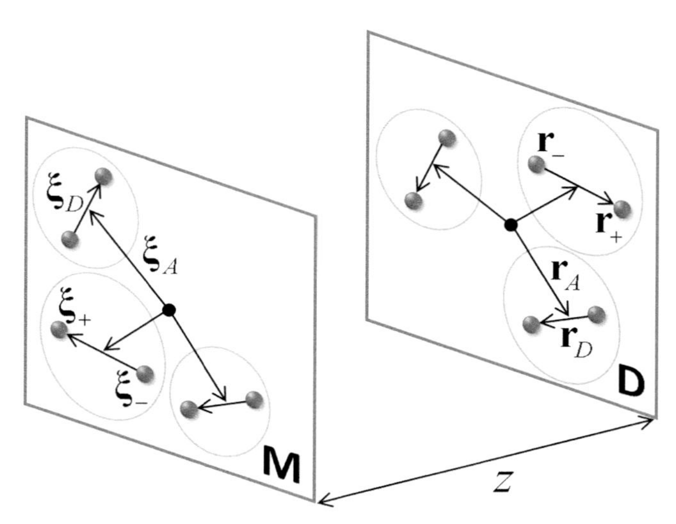

Fig. 2 Reduced coordinates for determining pairs of points in M and D. Shadowed circles depict non-locality supports at each plane, which enclose pairs of points of coordinates $\boldsymbol{\xi}_{\pm} = \boldsymbol{\xi}_A \pm \boldsymbol{\xi}_D/2$ on M and $\mathbf{r}_{\pm} = \mathbf{r}_A \pm \mathbf{r}_D/2$ on D.

By regarding Eq. (5), $|\mathbf{E}_n(\mathbf{r}, t)|^2$ implies non-locality conditions at M. This should be formalized by considering pairs of points $\boldsymbol{\xi}_{\pm} = \boldsymbol{\xi}_A \pm \boldsymbol{\xi}_D/2$ at M and $\mathbf{r}_{\pm} = \mathbf{r}_A \pm \mathbf{r}_D/2$ at D, where $(\boldsymbol{\xi}_A, \ \boldsymbol{\xi}_D)$ and $(\mathbf{r}_A, \ \mathbf{r}_D)$ denote reduced coordinates, with coordinate suffixed A specifying the midpoint between the considered pair and coordinate suffixed D representing the separation vector between them, as illustrated in Fig. 2 (Castañeda et al., 2020).

The analysis above leads straightforwardly to

$$\begin{aligned} H(\mathbf{r}_A) = \ & \frac{1}{2}\sum_n \hbar\omega_n \left(\hat{a}_n^\dagger \ \hat{a}_n + \hat{a}_n \ \hat{a}_n^\dagger\right) \\ & \int_M \int_M d^2\xi_A \ d^2\xi_D \ w_n(\boldsymbol{\xi}_+, \boldsymbol{\xi}_-) \ \tau(\boldsymbol{\xi}_+, \boldsymbol{\xi}_-) \ \mathbf{e}_n(\boldsymbol{\xi}_+) \\ & \cdot \ \mathbf{e}_n(\boldsymbol{\xi}_-) \ \Phi_n(\boldsymbol{\xi}_+, \boldsymbol{\xi}_-, \mathbf{r}_A, \mathbf{z}, k_n), \end{aligned} \tag{6}$$

that defines the quantum non-paraxial Hamiltonian for single photons, with $w_n(\boldsymbol{\xi}_+, \boldsymbol{\xi}_-) = \psi_n(\boldsymbol{\xi}_+) \ \psi_n^*(\boldsymbol{\xi}_-) = w_n^*(\boldsymbol{\xi}_-, \boldsymbol{\xi}_+)$ the prepared non-locality at M, $\tau(\boldsymbol{\xi}_+, \boldsymbol{\xi}_-) = t(\boldsymbol{\xi}_+) \ t^*(\boldsymbol{\xi}_-) = \tau^*(\boldsymbol{\xi}_-, \boldsymbol{\xi}_+)$ the non-local transmission at this plane, and

$$\begin{aligned}
&\Phi_n(\boldsymbol{\xi}_+, \boldsymbol{\xi}_-, \mathbf{r}_+, \mathbf{r}_-, \mathbf{z}, k_n) \\
&= \Theta_n(\boldsymbol{\xi}_+, \mathbf{r}_+, \mathbf{z}, k_n)\,\Theta_n^*(\boldsymbol{\xi}_-, \mathbf{r}_-, \mathbf{z}, k_n) \\
&= \left(\frac{k_n}{4\pi}\right)^2 \left(\frac{z + |\mathbf{z} + \mathbf{r}_+ - \boldsymbol{\xi}_+|}{|\mathbf{z} + \mathbf{r}_+ - \boldsymbol{\xi}_+|^2}\right)\left(\frac{z + |\mathbf{z} + \mathbf{r}_- - \boldsymbol{\xi}_-|}{|\mathbf{z} + \mathbf{r}_- - \boldsymbol{\xi}_-|^2}\right) \\
&\quad \exp(ik_n|\mathbf{z} + \mathbf{r}_+ - \boldsymbol{\xi}_+| - ik_n|\mathbf{z} + \mathbf{r}_- - \boldsymbol{\xi}_-|)
\end{aligned} \tag{7}$$

the second-order modes defined in the volume delimited by M and D, Fig. 2, with $\Phi_n(\boldsymbol{\xi}_+, \boldsymbol{\xi}_-, \mathbf{r}_+, \mathbf{r}_-, \mathbf{z}, k_n) = \Phi_n^*(\boldsymbol{\xi}_-, \boldsymbol{\xi}_+, \mathbf{r}_-, \mathbf{r}_+, \mathbf{z}, k_n)$. The kernel in Eq. (6) is given by the local component at D (i.e. for $\mathbf{r}_D = 0$) of Eq. (7), $\Phi_n(\boldsymbol{\xi}_+, \boldsymbol{\xi}_-, \mathbf{r}_A, \mathbf{z}, k_n) = \Theta_n(\boldsymbol{\xi}_+, \mathbf{r}_A, \mathbf{z}, k_n)\ \Theta_n^*(\boldsymbol{\xi}_-, \mathbf{r}_A, \mathbf{z}, k_n)$. The commutation rules of the creation and annihilation operators yield $\hat{a}_n^\dagger \hat{a}_n + \hat{a}_n\ \hat{a}_n^\dagger = 2\hat{a}_n^\dagger \hat{a}_n + 1$. Thus, the quantum non-paraxial Hamiltonian for single photons becomes

$$\begin{aligned}
H(\mathbf{r}_A) = \sum_n \hbar\omega_n \left(\hat{a}_n^\dagger \hat{a}_n + \frac{1}{2}\right) \int_M \int_M d^2\xi_A\ d^2\xi_D\ w_n(\boldsymbol{\xi}_+, \boldsymbol{\xi}_-)\ \tau(\boldsymbol{\xi}_+, \boldsymbol{\xi}_-) \\
\mathbf{e}_n(\boldsymbol{\xi}_+)\cdot\mathbf{e}_n(\boldsymbol{\xi}_-)\ \Phi_n(\boldsymbol{\xi}_+, \boldsymbol{\xi}_-, \mathbf{r}_A, \mathbf{z}, k_n).
\end{aligned} \tag{8}$$

The difference between the Hamiltonian in Eq. (8) and the standard paraxial Hamiltonian $H = \sum_n \hbar\omega_n\left(\hat{a}_n^\dagger\ \hat{a}_n + \frac{1}{2}\right)$ of quantum optics is just the spatial component, by which the non-paraxial Hamiltonian explicitly depends on the points $\mathbf{r}_A$ of D. This component gives new insight on the single photon behavior in the experimental setup, because it involves the setup configuration in the boundary conditions, which is completely absent in the standard paraxial quantum Hamiltonian. In contrast with this Hamiltonian, the non-paraxial quantum Hamiltonian seems to provide an accurate description of the spatial energy distribution of experimental outcomes. More precisely, its term for $\omega_n = \omega$,

$$\begin{aligned}
H(\mathbf{r}_A) = \hbar\omega \left(\hat{n} + \frac{1}{2}\right) \int_M \int_M d^2\xi_A\ d^2\xi_D \\
\kappa(\boldsymbol{\xi}_+, \boldsymbol{\xi}_-)\ \Phi(\boldsymbol{\xi}_+, \boldsymbol{\xi}_-, \mathbf{r}_A, \mathbf{z}, k),
\end{aligned} \tag{9}$$

with the non-locality function at M

$$\kappa(\boldsymbol{\xi}_+, \boldsymbol{\xi}_-) = w(\boldsymbol{\xi}_+, \boldsymbol{\xi}_-)\ \tau(\boldsymbol{\xi}_+, \boldsymbol{\xi}_-)\ \ \mathbf{e}(\boldsymbol{\xi}_+)\cdot\mathbf{e}(\boldsymbol{\xi}_-), \tag{10}$$

describes any individual experimental realization of single photon experiments appropriately. The crucial role of the non-locality as modal filter at

M is remarkable, mainly considering that the non-paraxial quantum Hamiltonian in Eq. (9) specifies the local values of energy at D, whose measurement ensures single photon detections at this plane.

2.2 Polarization and non-locality

The factor $\mathbf{e}(\boldsymbol{\xi}_+)\cdot\mathbf{e}(\boldsymbol{\xi}_-)$ of the non-locality function in Eq. (10) points out non-locality effects on the polarization state of the single photon that crosses a given point of M, in an individual experimental realization. These effects become evident when the crossing point conforms pairs with other points that belong to specific non-locality supports. For mathematical simplicity and without lack of generality, let us consider a double pinhole mask placed at M, with separation vector $\boldsymbol{\xi}_D$, as illustrated in Fig. 3. The non-locality support centered at $\boldsymbol{\xi}_A$ encloses the pinhole pair, and linear polarizers are inserted in the pinholes with their transmission and extinction axes oriented along the mutually orthogonal unit vectors $\{\mathbf{e}_{\|}(\boldsymbol{\xi}_\pm), \mathbf{e}_{\perp}(\boldsymbol{\xi}_\pm)\}$, respectively.

Let us assume that the single photon effectively arrives to the pinhole at $\boldsymbol{\xi}_+$, with the polarization state denoted by the unit vector $\mathbf{e}'(\boldsymbol{\xi}_+)$. The photon arrival to this pinhole is only restricted by the uncertainty principle. It is well-known that it can emerge with a new polarization state only denoted by $\mathbf{e}_{\|}(\boldsymbol{\xi}_+)$. The non-locality effect on polarization is configured by regarding the potential polarization state $\mathbf{e}_{\|}(\boldsymbol{\xi}_-)$ for the emerging single photon in the eventual case of crossing by $\boldsymbol{\xi}_-$. This consideration is not equivalent to the hypothesis of single photon delocalization. Instead of that, it fulfils the necessary condition for the non-local effects that both pinholes are transparent in some extent at the time the single photon crosses only one of them. Therefore $\mathbf{e}(\boldsymbol{\xi}_+)\cdot\mathbf{e}(\boldsymbol{\xi}_-) \equiv \mathbf{e}_{\|}(\boldsymbol{\xi}_+)\cdot\mathbf{e}_{\|}(\boldsymbol{\xi}_-)$ in Eq. (10).

Because $\mathbf{e}_{\|}(\boldsymbol{\xi}_\pm)$ are arbitrarily oriented on M, it is useful to express them on a cartesian fixed base of unit vectors $\{\mathbf{u}_x, \mathbf{u}_y\}$, Fig. 3. It should be noted that $\theta(\boldsymbol{\xi}_\pm) = \theta_\pm$ are the angles between $\mathbf{e}_{\|}(\boldsymbol{\xi}_\pm)$ and $\mathbf{u}_x$, while $\beta(\boldsymbol{\xi}_\pm) = \beta_\pm$ are the angles between $\mathbf{e}_{\|}(\boldsymbol{\xi}_\pm)$ and $\mathbf{e}'(\boldsymbol{\xi}_\pm)$.

In Dirac's notation, the unit vector bases $\{\mathbf{e}_{\|}(\boldsymbol{\xi}_\pm), \mathbf{e}_{\perp}(\boldsymbol{\xi}_\pm)\}$ and $\{\mathbf{u}_x, \mathbf{u}_y\}$ are represented by the orthonormal ket bases $\{|\|_\pm\rangle, |\perp_\pm\rangle\}$ and $\{|H\rangle, |V\rangle\}$ respectively, and the polarization states are denoted as $|p'(\boldsymbol{\xi}_+)\rangle = |p'_+\rangle$ for the incident single photon and $|p(\boldsymbol{\xi}_\pm)\rangle = |p_\pm\rangle$ for both the effectively and the potentially transmitted single photon. Therefore,

$$|p'_+\rangle = |\|_+\rangle\cos(\beta(\boldsymbol{\xi}_+)) + |\perp_+\rangle\sin(\beta(\boldsymbol{\xi}_+)) \tag{11}$$

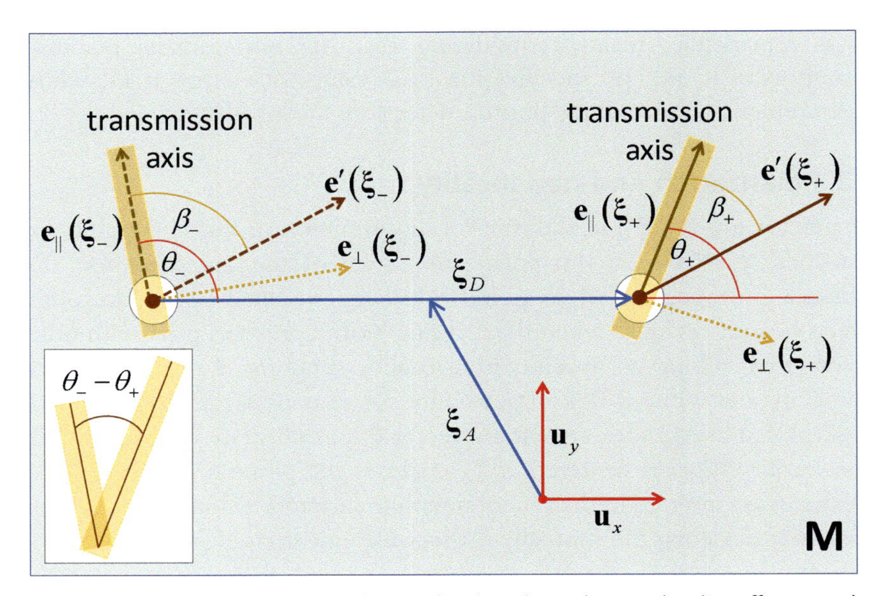

Fig. 3 Double pinhole mask placed at M for describing the non-locality effects on the single photon polarization state. Linear polarizers are inserted in the pinholes at the points $\xi_{\pm}$of the mask, whose transmission and extinction axes are represented by the mutually orthogonal unit vectors $\{\mathbf{e}_{\|}(\xi_{\pm}), \mathbf{e}_{\perp}(\xi_{\pm})\}$ respectively. $\beta_{\pm} = \beta(\xi_{\pm})$ and $\theta_{\pm} = \theta(\xi_{\pm})$. Solid line arrow for $\mathbf{e}'(\xi_{+})$ and parallel dotted line arrow for $\mathbf{e}'(\xi_{-})$ represent the individual experimental realization in which the single photon crosses only the pinhole at ξ_{+}.

describes the polarization state of the incident single photon to the pinhole at $\boldsymbol{\xi}_{+}$, while

$$|p_{+}\rangle = |\,\|_{+}\rangle \cos(\beta(\boldsymbol{\xi}_{+})) \tag{12a}$$

and

$$|p_{-}\rangle = |\,\|_{-}\rangle \cos(\beta(\boldsymbol{\xi}_{-})) \tag{12b}$$

represent the effective polarization state of the photon transmitted by the linear polarized at $\boldsymbol{\xi}_{+}$, and the potential polarization state to be transmitted by the linear polarized at $\boldsymbol{\xi}_{-}$ respectively. Therefore,

$$\langle p_{+}|p_{+}\rangle = \cos^{2}(\beta(\boldsymbol{\xi}_{+})) \tag{13a}$$

and

$$\langle p_{-}|p_{-}\rangle = \cos^{2}(\beta(\boldsymbol{\xi}_{-})) \tag{13b}$$

determine the quantum probabilities for the single photon to be transmitted by the linear polarizers at $\boldsymbol{\xi}_{\pm}$. Indeed, if the polarization state of the incident single photon is parallel to the transmission axes of the linear polarizers in the pinholes, $\beta(\xi_{\pm}) = 0, \ \pi$, then $\langle p_{\pm} | p_{\pm} \rangle = 1$ and the polarizers are completely transparent for the single photon, so that there is total certainty of its transmission. In contrast, if the polarization state of the incident single photon is orthogonal to the transmission axes of the linear polarizers, $\beta(\xi_{\pm}) = \pm \ \pi/2$, then $\langle p_{\pm} | p_{\pm} \rangle = 0$ and the polarizers are completely opaque for the single photon. Equivalently, $\langle a_+ | a_+ \rangle = \ \sin^2(\beta(\xi_+)) = 1$, with $| a_+ \rangle = \ | \perp_+ \rangle \sin(\beta(\xi_+))$, so that

$$\langle a_+ | a_+ \rangle = \ \sin^2(\beta(\xi_+)) \tag{14}$$

represents the quantum probability of single photon annihilation by the linear polarizer at ξ_+. Thus, there is total certainty of the single photon annihilation by $\beta(\xi_{\pm}) = \pm \ \pi/2$.

Because $\langle \perp_+ | \|_+ \rangle = \langle \|_+ | \perp_+ \rangle = 0$ then $\langle p_+ | a_+ \rangle = \langle a_+ | p_+ \rangle = 0$. Consequently,

$$\begin{aligned} \langle p'_+ | p'_+ \rangle &= \langle p_+ | p_+ \rangle + \langle a_+ | a_+ \rangle \\ &= \cos^2(\beta(\xi_+)) + \sin^2(\beta(\xi_+)) = 1. \end{aligned} \tag{15}$$

Eq. (15) points out that the transmission and annihilation of the single photon by the linear polarizer at ξ_+ are exclusive, exhaustive and complementary. Such processes determine a statistical bimodal distribution determined by Eqs. (13a), (14), that ensures the corpuscular nature of the single photon at M. In other words, each single photon should be completely transmitted or annihilated by the linear polarizer in the pinhole.

Now, let us consider that (Fig. 3)

$$| \|_{\pm} \rangle = \ | H \rangle \cos(\theta(\xi_{\pm})) + \ | V \rangle \sin(\theta(\xi_{\pm})). \tag{16}$$

Therefore, Ecs. (12 a, b) lead straightforwardly to

$$\begin{aligned} \langle p_- | p_+ \rangle = \ &\cos(\beta(\xi_+)) \ \cos(\beta(\xi_-)) \ (\cos(\theta(\xi_+))\cos(\theta(\xi_-)) \\ &+ \ \sin(\theta(\xi_+)) \ \sin(\theta(\xi_-))) \\ &= \cos(\theta(\xi_+) - \theta(\xi_-))\cos(\beta(\xi_+)) \ \cos(\beta(\xi_-)) \end{aligned} \tag{17}$$

that denotes the non-local link between the two pinholes, due to the linear polarizer transmissions. Indeed, after ensuring non-null transmissions of the linear polarizers at both pinholes, so that $\cos(\beta(\xi_+)) \ \cos(\beta(\xi_-)) \neq 0$, then $\langle p_- | p_+ \rangle = 0$ if the transmission axes of the polarizers are mutually orthogonal, i.e. $\theta(\xi_+) - \theta(\xi_-) = \pm\pi/2$, and $\langle p_- | p_+ \rangle = \pm\cos^2\beta$ if the transmission axes of the polarizers are mutually parallel, i.e. $\theta(\xi_+) - \theta(\xi_-) = 0,\ \pi$ and therefore $\beta(\xi_+) = \beta(\xi_-) = \beta$.

In addition, $\langle p_- | p_+ \rangle$ represents the factor $\mathbf{e}_{\|}(\xi_+)\cdot\mathbf{e}_{\|}(\xi_-) \equiv \mathbf{e}(\xi_+)\cdot\mathbf{e}(\xi_-)$ in the non-locality function in Eq. (10), which becomes

$$\kappa(\xi_+, \xi_-) = w(\xi_+, \xi_-)\ \tau(\xi_+, \xi_-)\ \cos(\theta(\xi_+) - \theta(\xi_-))\cos(\beta(\xi_+))\ \cos(\beta(\xi_-)), \tag{18}$$

whose local component (for $\xi_D = 0$)

$$\kappa(\xi_A, \xi_A) = |\psi(\xi_A)|^2\ |t(\xi_A)|^2\ \cos^2(\beta(\xi_A)) \tag{19}$$

determines the quantum probability for a single photon to cross the point ξ_A and to emerge with the polarization state $| \|_A \rangle$.

2.3 Single matter particles

In order to analyze the single matter particle propagation in the setup conceptually depicted in Fig. 1, let us start by considering the Schrödinger equation for field-free space, $\nabla^2\Psi(\mathbf{r}, t) - \frac{2m}{i\hbar}\frac{\partial\Psi(\mathbf{r}, t)}{\partial t} = 0$, whose solution is $\Psi_-(\mathbf{r}, t) = \psi(\mathbf{r})\ a\exp\left(-i\frac{E}{\hbar}t\right)$, where the scalar a is determined by initial conditions, and the spatial component $\psi(\mathbf{r})$ is the solution of the Helmholtz equation $(\nabla^2 + k^2)\psi(\mathbf{r}) = 0$, with $k = p/\hbar$. In these expressions, $m,\ E,\ p$ denotes the mass, the energy and the momentum of the particle. By following the same reasoning in Section 2.1, exact (nonparaxial) $\psi(\mathbf{r})$ takes the form

$$\psi(\mathbf{r}) = \int_M d^2\xi\ \psi(\xi)\ t(\xi)\ \Theta(\xi, \mathbf{r}, \mathbf{z}, k), \tag{20}$$

where $\psi(\xi)$ is the eigenfunction of the Laplacian at the M plane, $t(\xi)$ is the complex transmission function at such plane and the kernel is given by Eq. (3). By considering the quantum Hamiltonian $\hat{H}_- \equiv -\frac{\hbar}{i}\ \frac{\partial}{\partial t}$ and that $\Psi_-(\mathbf{r}, t)$ is the representation of ket $|\Psi_-\rangle$ in ordinary space-time, the matter particle energy recorded at the point $\mathbf{r}$ is given by the positive-definite quantity

$$\langle \Psi_- | \hat{H}_- | \Psi_- \rangle \equiv E\,|\psi(\mathbf{r})|^2\ a^*a, \tag{21}$$

where $|\psi(\mathbf{r})|^2$ denotes the quantum probability to find the particle at a given point and a^*a can be connected with the number of particles with energy E that arrive to the point $\mathbf{r}$.

It should be noted that $\Psi_-(\mathbf{r}, t)$ describes the direction of propagation corresponding to one of the degrees of freedom in the orientation of the experimental setup optical axis. The description independent of the experimental setup orientation should be obtained by including the degree of freedom corresponding to the opposite direction of propagation, which is described by $\Psi_+(\mathbf{r}, t) = \psi(\mathbf{r})\ a^*\exp\left(i\frac{E}{\hbar}t\right)$, with the energy given by $E\,|\psi(\mathbf{r})|^2\ \ a\,a^*$. Two quantum features should be ensured to perform it consistently, i.e. non-commutativity and time odd symmetry. The first feature concerns the order of the products $a\,a^*$ in the last expression and a^*a in Eq. (21), which should be explicitly maintained to associate a and a^* appropriately to non-commutative operators related to the number operator. The second feature is implicit in Schrödinger equation, i.e. the sign inversion in the exponential argument of $\Psi_+(\mathbf{r}, t)$ with respect to $\Psi_-(\mathbf{r}, t)$ involves time reversal in Schrödinger equation. Indeed, Helmholtz equation $(\nabla^2 + k^2)\psi(\mathbf{r}) = 0$, with $k = p/\hbar = \sqrt{2mE}/\hbar$, leads to $\nabla^2\Psi_+(\mathbf{r}, t) = -k^2\Psi_+(\mathbf{r}, t) = -\frac{2mE}{\hbar^2}\Psi_+(\mathbf{r}, t)$, while $\frac{2m}{i\hbar}\frac{\partial\Psi_+(\mathbf{r}, t)}{\partial t} = \frac{2mE}{\hbar^2}\Psi_+(\mathbf{r}, t)$. This means that $\nabla^2\Psi(\mathbf{r}, t) + \frac{2m}{i\hbar}\frac{\partial\Psi(\mathbf{r}, t)}{\partial t} = 0$ which is achieved by introducing time reversal in Schrödinger equation, and therefore in the Hamiltonian, i.e. $\hat{H}_+ \equiv \frac{\hbar}{i}\frac{\partial}{\partial t}$. Accordingly,

$$\langle\Psi_+|\hat{H}_+|\Psi_+\rangle \equiv E\,|\psi(\mathbf{r})|^2\ \ a\,a^*. \tag{22}$$

Therefore, the single matter particle energy recorded at the point $\mathbf{r}$ with independence of the experimental setup orientation should be expressed as

$$\frac{1}{2}\left(\langle\Psi_-|\hat{H}_-|\Psi_-\rangle + \langle\Psi_+|\hat{H}_+|\Psi_+\rangle\right) \equiv \frac{E}{2}\ \ |\psi(\mathbf{r})|^2\ \ (a^*a + a\,a^*). \tag{23}$$

Eq. (20) and the reduced coordinates depicted in Fig. 2 allow expressing Eq. (23) as

$$H(\mathbf{r}_A) = \frac{E}{2}\ (\hat{a}^\dagger\hat{a} + \hat{a}\,\hat{a}^\dagger)\int_M\int_M d^2\xi_A\ d^2\xi_D\ \kappa(\boldsymbol{\xi}_+, \boldsymbol{\xi}_-)\ \Phi(\boldsymbol{\xi}_+, \boldsymbol{\xi}_-, \mathbf{r}_A, \mathbf{z}, k), \tag{24}$$

with $\frac{1}{2}\left(\langle\Psi_-|\hat{H}_-|\Psi_-\rangle + \langle\Psi_+|\hat{H}_+|\Psi_+\rangle\right) \equiv H(\mathbf{r}_A)$ and

$$\kappa(\boldsymbol{\xi}_+, \boldsymbol{\xi}_-) = w(\boldsymbol{\xi}_+, \boldsymbol{\xi}_-)\ \tau(\boldsymbol{\xi}_+, \boldsymbol{\xi}_-) \tag{25}$$

the non-locality function at M, where $w(\xi_+, \xi_-) = \psi(\xi_+)\ \psi^*(\xi_-) = w^*(\xi_+, \xi_-)$ denotes the prepared non-locality and $\tau(\xi_+, \xi_-) = t(\xi_+)\ t^*(\xi_-) = \tau^*(\xi_+, \xi_-)$ denotes the non-local transmission at this plane. The non-local kernel $\Phi(\xi_+, \xi_-, \mathbf{r}_A, \mathbf{z}, k) = \Phi^*(\xi_-, \xi_+, \mathbf{r}_A, \mathbf{z}, k)$ specifies the second-order modes defined in the volume delimited by M and D, Fig. 2, for $\mathbf{r}_D = 0$ in any individual experimental realization with a single matter particle, which is a mono-energetic event with the particle momentum $p = \sqrt{2mE}$ only restricted by the uncertainty principle $\Delta x\ \Delta p \geq \hbar/2$. It should be kept in mind that only a single matter particle moves in the setup in any individual experimental realization, without interacting in any way with the matter particles in the preceding and the posterior individual realizations, because it is created by a source emission event and it is annihilated by a detector recording event. It allows considering the scalars a^*, a as eigenvalues of the creation and annihilation operators $\hat{a}^\dagger$, $\hat{a}$ and replacing them correspondently, as expressed in Eq. (24).

This remains valid for quasi-monoenergetic and poly-energetic beams of matter particles, even though most of matter particles are fermions that should obey the exclusion principle (Feynman et al., 1965). Indeed, the set of individual experimental realizations is a binary sequence of events of value 1 (only one matter particle systems), followed by events of value 0 (systems with null matter particles). Particles in different individual realizations can exhibit different momenta in case of experiments with poly-energetic beams. Nevertheless, the complete experimental outcomes in all the cases is achieved by the accumulation of the local detection of the single matter particle in each individual realization.

Therefore, single particle experiments clearly differ from many-particle systems, in which a set of fermions should be distributed, at a time, in a set of states in accordance with the exclusion principle, in such a way that the anti-commuting rule $\{\hat{a}^\dagger, \hat{a}\} = \hat{a}^\dagger\hat{a} + \hat{a}\ \hat{a}^\dagger = 1$ should be fulfilled (Zeidler, 2011). In order to establish the commutation rule between creation and annihilation operators of single matter particles in any individual experimental realization, let us denote $\omega = E/\hbar$ and introduce the canonical variables (Zeidler, 2011) for the single matter particle position

$$q(t) = a^*\exp(i\omega t) + a\exp(-i\omega t), \tag{26}$$

and momentum $p(t) = \frac{\partial q(t)}{\partial t}$ in the setup, i.e.

$$p(t) = -i\omega(-a^{*}\exp(i\omega t) + a\exp(-i\omega t)), \tag{27}$$

so that $-\omega^2 q(t) = \frac{\partial p(t)}{\partial t}$. By following the standard quantum procedure, these canonical variables can be replaced by operators $\hat{q}$ and $\hat{p}$ respectively, with commutation rules $[\hat{q}, \hat{p}] = i\hbar$ and $[\hat{q}, \hat{q}] = [\hat{p}, \hat{p}] = 0$. In turn, such operators allow define the annihilation

$$\hat{a} = \frac{1}{\sqrt{2\hbar\omega}}(\omega\hat{q} + i\hat{p}), \tag{28}$$

and creation

$$\hat{a}^{\dagger} = \frac{1}{\sqrt{2\hbar\omega}}(\omega\hat{q} - i\hat{p}), \tag{29}$$

operators, with commutation rules $[\hat{a}, \hat{a}^{\dagger}] = 1$ and $[\hat{a}, \hat{a}] = [\hat{a}^{\dagger}, \hat{a}^{\dagger}] = 0$. Accordingly, $\hat{a}^{\dagger}\hat{a} + \hat{a}\,\hat{a}^{\dagger} = 2\hat{a}^{\dagger}\hat{a} + 1$ in Eq. (24), with $\hat{a}^{\dagger}\hat{a} = \hat{n}$ the number operator. So, Eq. (24) becomes

$$\begin{aligned} H(\mathbf{r}_A) = \hbar\omega\left(\hat{n} + \frac{1}{2}\right) \int_M \int_M d^2\xi_A \; d^2\xi_D \\ \kappa(\boldsymbol{\xi}_+, \boldsymbol{\xi}_-) \; \Phi(\boldsymbol{\xi}_+, \boldsymbol{\xi}_-, \mathbf{r}_A, \mathbf{z}, k). \end{aligned} \tag{30}$$

with $\omega = E/\hbar$ (i.e. the ratio, with frequency units, between the single matter particle energy and the universal constant $\hbar$).

It is worth noting that Eqs. (9) and (30) exhibit the same mathematical form with the same physical meaning, thus supporting the feasibility of a unified phenomenology to describe the propagation in ordinary space of quantum corpuscular entities as single photons and single matter particles. Specifically, such equations describe the spatial distribution of the energy of n corpuscular entities on non-paraxial propagation in a sequence of individual experimental realizations. This generalized description is exact and stationary (time-independent) but explicitly depends on the non-paraxial, non-local modes and the number of photons or matter particles, whose fundamental energy (i.e. the energy of a single photon or single matter particle) is expressed by the quantum formula $E = \hbar\omega$. For this reason, we call any of Eqs. (9) and (30) the *canonical quantum equation* of non-paraxial propagation in ordinary space.

3. The quantum interference operator

The canonical quantum equation can be obtained in terms of the product of functions of the Hilbert space, that constitute the coordinate representation of the corresponding kets in Dirac notation (Mandel & Wolf, 1995). Indeed,

$$H(\mathbf{r}_A) = \Upsilon(\mathbf{r}_A)\ \Upsilon^*(\mathbf{r}_A) \equiv \langle \mathbf{r}_A | \Upsilon \rangle \langle \Upsilon | \mathbf{r}_A \rangle \tag{31}$$

because it takes on positive real values at each point $\mathbf{r}_A$ of D. In addition, the kernel $\Phi(\boldsymbol{\xi}_+, \boldsymbol{\xi}_-, \mathbf{r}_A, \mathbf{z}, k) = \Theta(\boldsymbol{\xi}_+, \mathbf{r}_A, \mathbf{z}, k)\quad \Theta^*(\boldsymbol{\xi}_-, \mathbf{r}_A, \mathbf{z}, k)$ can be expressed as $\Phi(\boldsymbol{\xi}_+, \boldsymbol{\xi}_-, \mathbf{r}_A, \mathbf{z}, k) \equiv \langle \mathbf{r}_A | \Theta(\boldsymbol{\xi}_+) \rangle \langle \Theta(\boldsymbol{\xi}_-) | \mathbf{r}_A \rangle$. Accordingly,

$$|\psi(\mathbf{r}_A)|^2 \equiv \langle \mathbf{r}_A | \psi \rangle \langle \psi | \mathbf{r}_A \rangle = \int_M \int_M d^2\xi_A\ d^2\xi_D\ \kappa(\boldsymbol{\xi}_+, \boldsymbol{\xi}_-)\ \langle \mathbf{r}_A | \Theta(\boldsymbol{\xi}_+) \rangle \langle \Theta(\boldsymbol{\xi}_-) | \mathbf{r}_A \rangle. \tag{32}$$

Eqs. (31) and (32) lead to the definition of the self-adjoin density operator

$$\begin{aligned} |\Upsilon\rangle\langle\Upsilon| &= \hbar\omega \left(\hat{n} + \frac{1}{2}\right) |\psi\rangle\langle\psi| \\ &= \hbar\omega \left(\hat{n} + \frac{1}{2}\right) \int_M \int_M d^2\xi_A\ d^2\xi_D \\ &\quad \kappa(\boldsymbol{\xi}_+, \boldsymbol{\xi}_-) |\Theta(\boldsymbol{\xi}_+)\rangle\langle\Theta(\boldsymbol{\xi}_-)|, \end{aligned} \tag{33}$$

whose coordinate representation, Ec. (31), is the canonical quantum equation.

3.1 The role of vacuum and the geometric states of space

The canonical quantum equation for n single photons or single matter particles is obtained from the density operator in Eq. (33) as

$$\begin{aligned} \langle n, \mathbf{r}_A | \Upsilon \rangle \langle \Upsilon | n, \mathbf{r}_A \rangle &= \hbar\omega \left(\langle n | \hat{n} | n \rangle + \tfrac{1}{2} \right) \langle \mathbf{r}_A | \psi \rangle \langle \psi | \mathbf{r}_A \rangle \\ = \hbar\omega \left(n + \tfrac{1}{2} \right) & \int_M \int_M d^2\xi_A\ d^2\xi_D\ \kappa(\boldsymbol{\xi}_+, \boldsymbol{\xi}_-) \langle \mathbf{r}_A | \Theta(\boldsymbol{\xi}_+) \rangle \langle \Theta(\boldsymbol{\xi}_-) | \mathbf{r}_A \rangle \end{aligned} \tag{34}$$

A novel and especially important phenomenological implication of Eq. (34) concerns its form in vacuum, i.e. in the absence of photons or matter particles in the setup that characterizes the events of value 0 for instance. Vacuum is therefore denoted by $\langle 0|\hat{n}|0\rangle = 0$, so that Eq. (33) or equivalently Eq. (34) gives the density operator for vacuum

$$\begin{aligned} \langle 0|\Upsilon\rangle\langle\Upsilon|0\rangle &= |\Upsilon_0\rangle\langle\Upsilon_0| = \frac{\hbar\omega}{2} \\ |\psi\rangle\langle\psi| &= \frac{\hbar\omega}{2} \int_M \int_M d^2\xi_A \; d^2\xi_D \;\; \kappa(\boldsymbol{\xi}_+, \boldsymbol{\xi}_-) \, |\Theta(\boldsymbol{\xi}_+)\rangle\langle\Theta(\boldsymbol{\xi}_-)| \,. \end{aligned} \tag{35}$$

This surprising result points out that the non-paraxial, non-local geometric modes, specified by the self-adjoin density matrix $|\Theta(\boldsymbol{\xi}_+)\rangle\langle\Theta(\boldsymbol{\xi}_-)|$ can live in vacuum with the vacuum energy $\hbar\omega/2$. In other words, their existence seems not to require the presence of photons or particles in the setup. This allows us to propose a radical change of the Newtonian conception of ordinary space. Instead of a uniform, isotropic, non-deformable and passive distribution of points, ordinary space should be conceived as an infinite set of geometric modes with the vacuum energy, given by the density matrix $\frac{\hbar\omega}{2}|\Theta(\boldsymbol{\xi}_+)\rangle\langle\Theta(\boldsymbol{\xi}_-)|$. We call them *the geometric states of space.*

The non-locality function $\kappa(\boldsymbol{\xi}_+, \boldsymbol{\xi}_-)$ in Eq. (35) behaves as a modal filter that selects and weights the specific finite subset of geometric states of space that characterizes each experimental setup. The filtering is performed by the prepared non-locality at M, $w(\boldsymbol{\xi}_+, \boldsymbol{\xi}_-)$, and the effective non-local transmission at this plane, $\tau_{eff}(\boldsymbol{\xi}_+, \boldsymbol{\xi}_-)$, that can include polarization components in case of single photon experiments. This setup role, clearly established in vacuum, is consistent with the fundamental quantum mechanical ideas early discussed by Bohr (Bohr, 1935).

In order to interpret the geometric states of space phenomenologically, it is useful to separate the contributions provided to the density operator for vacuum by single points and by pairs of points of M. It is straightforwardly performed by inserting the dimensionless function $1 \equiv \delta(\boldsymbol{\xi}_D) + (1 - \delta(\boldsymbol{\xi}_D))$, with $\delta(\boldsymbol{\xi}_D)$ the Dirac delta, in the integrand of Eq. (35). It gives,

$$\begin{aligned} |\Upsilon_0\rangle\langle\Upsilon_0| &= \frac{\hbar\omega}{2} \int_M d^2\xi_A \;\; \hat{W}(\boldsymbol{\xi}_A) \\ &= \frac{\hbar\omega}{2} \int_M d^2\xi_A \left(\hat{R}(\boldsymbol{\xi}_A) + \frac{1}{2}\hat{G}(\boldsymbol{\xi}_A)\right), \end{aligned} \tag{36}$$

where the self-adjoint density operators

$$\hat{R}(\boldsymbol{\xi}_A) = \kappa(\boldsymbol{\xi}_A, \boldsymbol{\xi}_A)\,|\Theta(\boldsymbol{\xi}_A)\rangle\langle\Theta(\boldsymbol{\xi}_A)| \tag{37}$$

and

$$\hat{G}(\xi_A) = \int_{\substack{M \\ \xi_D \neq 0}} d^2\xi_D \ \kappa(\xi_+, \xi_-) \,|\, \Theta(\xi_+) \rangle \langle \Theta(\xi_-) \,| \tag{38}$$

denote, respectively, the ground states of space in the volume delimited by M and D, and the geometric potential that excites them (Castañeda et al., 2020, 2023).

Eq. (37) points out that all the ground states of space have the same geometry. They are cones with vertices at the points ξ_A of M, bases covering D and Lorentzian cross-sections, described by the coordinate representation of the modes of the density matrix $|\,\Theta(\xi_A)\rangle\langle\Theta(\xi_A)\,|$,

$$\langle \mathbf{r}_A \,|\, \Theta(\xi_A) \rangle \langle \Theta(\xi_A) \,|\, \mathbf{r}_A \rangle = |\,\Theta(\xi_A, \mathbf{r}_A, \mathbf{z}, k)\,|^2 = \left(\frac{k}{4\pi}\right)^2 \left(\frac{z + |\mathbf{z} + \mathbf{r}_A - \xi_A|}{|\mathbf{z} + \mathbf{r}_A - \xi_A|^2}\right)^2. \tag{39}$$

The specific subset of ground states of space for a given setup is filtered by the local component of the non-locality function established in vacuum at M, $\kappa(\xi_A, \xi_A) = |\,\psi(\xi_A)\,|^2 \ |\,t_{eff}(\xi_A)\,|^2$, with $0 \leq |\,t_{eff}(\xi_A)\,|^2 \leq 1$ the effective transmittance of the device placed at M. In general, $|\,t_{eff}(\xi_A)\,|^2 = |\,t(\xi_A)\,|^2 \cos^2(\beta(\xi_A))$ for single photons and $|\,t_{eff}(\xi_A)\,|^2 = |\,t(\xi_A)\,|^2$ for single matter particles. Therefore, $\kappa(\xi_A, \xi_A)$ also determines the accessibility to the filtered ground states by means of the values of the prepared quantum probability and the effective transmittance at the cone vertex, which is the point where a single photon or matter particle enters the ground state.

The contribution of each ground state of space to the density operator for vacuum in Eq. (36) takes the form $\frac{\hbar\omega}{2}\hat{R}(\xi_A) = \frac{\hbar\omega}{2}\kappa(\xi_A, \xi_A)\,|\,\Theta(\xi_A)\rangle\langle\Theta(\xi_A)\,|$, whose coordinate representation is

$$\frac{\hbar\omega}{2}\langle \mathbf{r}_A \,|\, \hat{R}(\xi_A) \,|\, \mathbf{r}_A \rangle = \frac{\hbar\omega}{2}\left(\frac{k}{4\pi}\right)^2 |\,\psi(\xi_A)\,|^2 \ |\,t_{eff}(\xi_A)\,|^2 \left(\frac{z + |\mathbf{z} + \mathbf{r}_A - \xi_A|}{|\mathbf{z} + \mathbf{r}_A - \xi_A|^2}\right)^2. \tag{40}$$

Eq. (40) points out that the Lorentzian cross-section of the ground state at D distributes the quantum probability with its maximum on the cone axis and a monotonic decay asymptotic to null with the distance to the axis. The quantum probability becomes negligible outside of the effective angular aperture of the cone (i.e. ca. 70° with respect to the cone axis for a decay of ca. 95 %) (Castañeda et al., 2020). Therefore, the ground state is actually a Lorentzian well that confines single photons and single matter particles to move preferably around the well axis.

In turn, each non-local mode of the geometric potential is associated to a specific pair of points $\boldsymbol{\xi}_{\pm}$ and provides the same spatial modulation to the ground states of space with vertices at such pair of points. Two excited states of space are thus producing at a time in the whole volume delimited by M and D, whose spatial modulations characterize the phenomenon called interference. It is worth noting that the ground states in Eq. (40) cannot excite each other geometrically as their Lorentzian wells overlap, Fig. 4 upper row. The excitation is exclusively provided by the geometric potential, which in turn is activated by the prepared non-locality at M, Fig. 4 middle row. It spatially modulates the quantum probability density on D, thus determining specific confinement zones, Fig. 4 bottom row. Therefore, in the absence of geometric potential, $\langle \mathbf{r}_A | \hat{R}(\boldsymbol{\xi}_A) | \mathbf{r}_A \rangle$ denotes a separable set of Lorentzian wells.

In this context, both the ground states of space and the geometric potential modes live in vacuum with the vacuum energy, in such a way that the excited states of space seem to be provided by vacuum in a non-Newtonian ordinary space. In other words, a generalized non-paraxial quantum theory of interference should attribute interference to the excited states of space in vacuum instead of the wave superposition. So, the interference pattern revealed by the detection of a large enough number of single photons or single matter particles is the map that they perform of the excited geometric states of space in the setup volume.

The excited geometric estate of space with vertex at a specific point $\boldsymbol{\xi}_A$ of M is described by the self-adjoint operator $\hat{W}(\boldsymbol{\xi}_A) = \hat{R}(\boldsymbol{\xi}_A) + \frac{1}{2}\hat{G}(\boldsymbol{\xi}_A)$. It is worth noting that ground states of space do no exhibit any spatial modulation in the absence of geometric potential modes. This means that the geometric potential is a necessary and sufficient condition for interference. Furthermore, Eq. (38) points out that only the non-local, non-paraxial modes for pairs of points in the non-locality supports (i.e. pairs of points at which $\kappa(\boldsymbol{\xi}_+, \boldsymbol{\xi}_-) \neq 0$) effectively contribute to the geometric potential. In other words, the filtering provided by the non-locality function $\kappa(\boldsymbol{\xi}_+, \boldsymbol{\xi}_-)$ activates the specific geometric potential in a given setup. Accordingly, interference does not occur if $\kappa(\boldsymbol{\xi}_+, \boldsymbol{\xi}_-) = 0$ for any $\boldsymbol{\xi}_D \neq 0$ because $\hat{G}(\boldsymbol{\xi}_A) = 0$ and $\hat{W}(\boldsymbol{\xi}_A) = \hat{R}(\boldsymbol{\xi}_A)$ in this case.

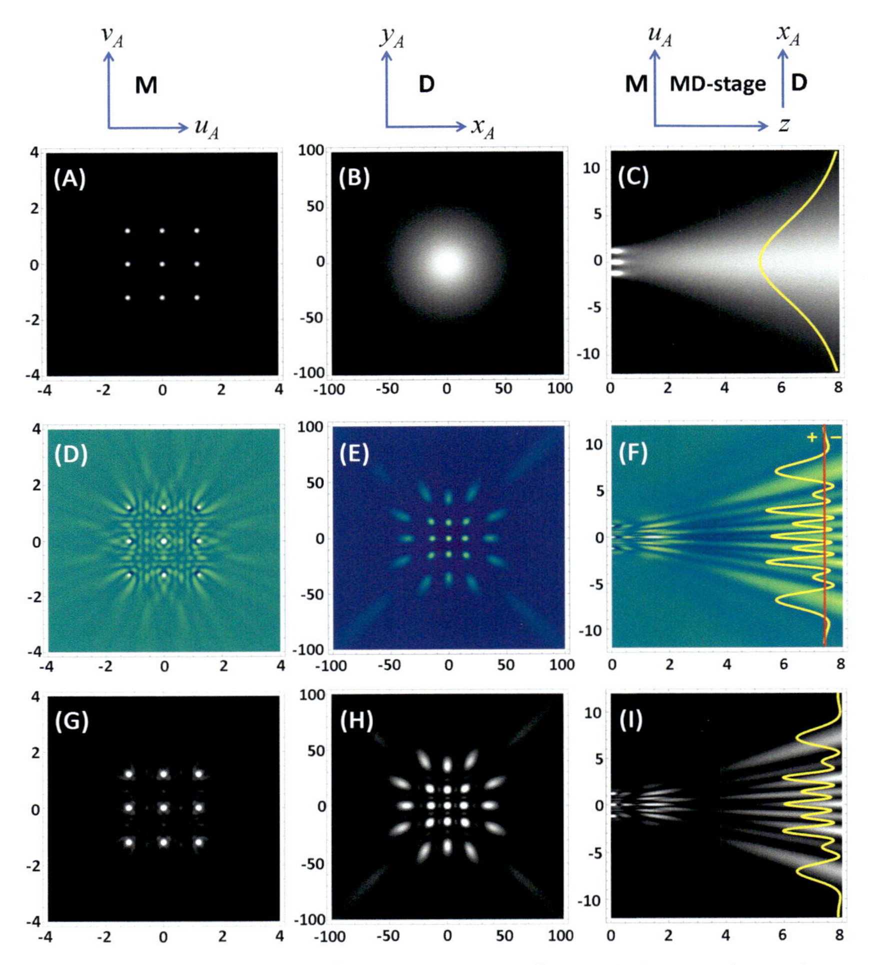

Fig. 4 Ground states and excited geometric states of space in the experimental setup depicted in Fig. 2, whose vertices distribute in an array of 3×3 points of M shown on the left column. The horizontal and vertical array spacing is $a = 3\lambda$ ($\lambda = 0.4\ \mu m$ for single photons and $\lambda = 0.4\ pm$ for single matter particles). Left column: cross sections at M with $\boldsymbol{\xi}_A \equiv (u_A, v_A)$. Middle column: cross sections at D ($z = 100\lambda$) with $\mathbf{r}_A \equiv (x_A, y_A)$. Right column: axial sections in MD-stage ($0 \leq z \leq 20\lambda$). (A)-(C) Lorentzian wells of the ground states. (D)-(F) Geometric potential activated by the highest prepared non-locality at M. (G)-(I) Spatially structured Lorentzian wells established by exciting the ground states on the upper row by the geometric potential on the middle row. The axes units are μm for single photons and pm for single matter particles.

The causal relationship between the non-locality function at M and the immediate activation of the non-local modes in the setup volume, implicit in the geometric potential, allows characterizing interference as a phenomenon with non-local causes (Brunner et al., 2014; Hall, 2016). Nevertheless, interference does not violate the limit for light signaling established in relativity (i.e. it is not a "spooky action at a distance") because the non-local, non-paraxial modes live in vacuum and are displayed in the full space. So, the setup delimits their volume segments between the planes M and D, with the vertices at M. Furthermore, $\kappa(\boldsymbol{\xi}_+, \boldsymbol{\xi}_-)$ does not contain creation operators for the modes but only components that select and weight passively the existing modes in the setup volume.

In order to describe the geometric excitations provided by the geometric potential, it is useful to express Eq. (38) by considering the two degrees in orientation of the separation vectors $\boldsymbol{\xi}_D$ and the hermitic symmetry of the non-locality function, $\kappa(\boldsymbol{\xi}_+, \boldsymbol{\xi}_-) = w(\boldsymbol{\xi}_+, \boldsymbol{\xi}_-)\ \tau_{eff}(\boldsymbol{\xi}_+, \boldsymbol{\xi}_-) = \kappa^*(\boldsymbol{\xi}_-, \boldsymbol{\xi}_+)$, with $w(\boldsymbol{\xi}_+, \boldsymbol{\xi}_-) = |w(\boldsymbol{\xi}_+, \boldsymbol{\xi}_-)|\exp(i\alpha(\boldsymbol{\xi}_+, \boldsymbol{\xi}_-))$, $\tau_{eff}(\boldsymbol{\xi}_+, \boldsymbol{\xi}_-) = |\tau_{eff}(\boldsymbol{\xi}_+, \boldsymbol{\xi}_-)|\exp(i\Delta\phi(\boldsymbol{\xi}_+, \boldsymbol{\xi}_-))$ and $\Delta\phi(\boldsymbol{\xi}_+, \boldsymbol{\xi}_-) = \phi(\boldsymbol{\xi}_+) - \phi(\boldsymbol{\xi}_-)$. This gives

$$\hat{G}(\boldsymbol{\xi}_A) = \int_{\boldsymbol{\xi}_D \neq 0} {}_M\, d^2\xi_D |\kappa(\boldsymbol{\xi}_+, \boldsymbol{\xi}_-)| (|\Theta(\boldsymbol{\xi}_+)\rangle\langle\Theta(\boldsymbol{\xi}_-)|\exp(i\alpha(\boldsymbol{\xi}_+, \boldsymbol{\xi}_-) + i\Delta\phi(\boldsymbol{\xi}_+, \boldsymbol{\xi}_-)) + |\Theta(\boldsymbol{\xi}_-)\rangle\langle\Theta(\boldsymbol{\xi}_+)|\exp(-i\alpha(\boldsymbol{\xi}_+, \boldsymbol{\xi}_-) - i\Delta\phi(\boldsymbol{\xi}_+, \boldsymbol{\xi}_-))) \tag{41}$$

whose coordinate representation yields

$$\langle \mathbf{r}_A | \hat{G}(\boldsymbol{\xi}_A) | \mathbf{r}_A \rangle = 2\left(\frac{k}{4\pi}\right)^2 \int_{\boldsymbol{\xi}_D \neq 0} {}_M\, d^2\xi_D |w(\boldsymbol{\xi}_+, \boldsymbol{\xi}_-)|\,|\tau_{eff}(\boldsymbol{\xi}_+, \boldsymbol{\xi}_-)| \left(\frac{z + |\mathbf{z} + \mathbf{r}_A - \boldsymbol{\xi}_+|}{|\mathbf{z} + \mathbf{r}_A - \boldsymbol{\xi}_+|^2}\right)\left(\frac{z + |\mathbf{z} + \mathbf{r}_A - \boldsymbol{\xi}_-|}{|\mathbf{z} + \mathbf{r}_A - \boldsymbol{\xi}_-|^2}\right) \times \cos(k|\mathbf{z} + \mathbf{r}_A - \boldsymbol{\xi}_+| - k|\mathbf{z} + \mathbf{r}_A - \boldsymbol{\xi}_-| + \alpha(\boldsymbol{\xi}_+, \boldsymbol{\xi}_-) + \Delta\phi(\boldsymbol{\xi}_+, \boldsymbol{\xi}_-)) \tag{42}$$

Therefore, the mode of the geometric potential associated to the pair of points $\boldsymbol{\xi}_\pm$ provides a cosine-like spatial modulation to the ground states of space with vertices at such points, only if such ground states are both accessible, i.e. if their vertices belong to a non-locality support of $w(\boldsymbol{\xi}_+, \boldsymbol{\xi}_-)$ and both vertices are transparent in some extension, so that $\tau_{eff}(\boldsymbol{\xi}_+, \boldsymbol{\xi}_-) \neq 0$. This geometric excitation distributes the quantum probability in specific zones within the volume of the ground states, as described by the coordinate representation of the density operator for vacuum in Eq. (36),

$$|\Upsilon_0(\mathbf{r}_A)|^2 = \langle \mathbf{r}_A | \Upsilon_0 \rangle \langle \Upsilon_0 | \mathbf{r}_A \rangle = \frac{\hbar\omega}{2} \int_M d^2\xi_A \ \langle \mathbf{r}_A | \hat{W}(\boldsymbol{\xi}_A) | \mathbf{r}_A \rangle. \quad (43)$$

As a consequence, spatially structured Lorentzian wells are established in the setup volume, characterized by specific distributions of confinement zones where the quantum probability grows. The distribution of the confinement zones in the cross-section of the spatially structured Lorentzian wells at D configures the interference pattern to be mapped by single photons or single matter particles.

However, the coordinate representation of each individual excited state of space $\langle \mathbf{r}_A | \hat{W}(\boldsymbol{\xi}_A) | \mathbf{r}_A \rangle = \langle \mathbf{r}_A | \hat{R}(\boldsymbol{\xi}_A) | \mathbf{r}_A \rangle + \frac{1}{2} \langle \mathbf{r}_A | \hat{G}(\boldsymbol{\xi}_A) | \mathbf{r}_A \rangle$ takes on negative values at the points $\mathbf{r}_A$ where $\langle \mathbf{r}_A | \hat{G}(\boldsymbol{\xi}_A) | \mathbf{r}_A \rangle < 0$ and $\langle \mathbf{r}_A | \hat{R}(\boldsymbol{\xi}_A) | \mathbf{r}_A \rangle < \frac{1}{2} |\langle \mathbf{r}_A | \hat{G}(\boldsymbol{\xi}_A) | \mathbf{r}_A \rangle|$. Therefore, such zones are non-accessible or forbidden for photons and matter particles that propagate within this individual excited state, because their negative values do not represent quantum probabilities of position. Nevertheless, the overlap of the set of excited states of space for a given setup must fulfil Eq. (43). This means that (i) the excited states of space constitute a non-separable set, and (ii) the negative values of the forbidden zones of each individual excited state must be removed by the confinement zones of the remainder excited states.

Because of the phenomenological description above, which is original of the proposed theory, we call the density operator for vacuum in Eq. (36) *the quantum non-paraxial operator for interference* in vacuum. This novel operator suggests the confinement in spatially structured Lorentzian wells as phenomenological principle for recording interference patterns, which is quite different of the wave superposition that supports the standard description of optical and quantum interference.

3.2 The prepared non-locality

As expressed before, the geometric potential seems to be the necessary and sufficient condition for interference. Its non-paraxial modes are filtered and activated by the non-locality function $\kappa(\boldsymbol{\xi}_+, \boldsymbol{\xi}_-) = w(\boldsymbol{\xi}_+, \boldsymbol{\xi}_-) \ \tau_{eff}(\boldsymbol{\xi}_+, \boldsymbol{\xi}_-)$, in such a way that the prepared non-locality $w(\boldsymbol{\xi}_+, \boldsymbol{\xi}_-)$ at M becomes of crucial importance in interference. Below, we show that $w(\boldsymbol{\xi}_+, \boldsymbol{\xi}_-)$ is prepared in vacuum, in a previous stage arranged to the setup in Fig. 2 by adding a third plane, at a distance z' from M. The effective source of single

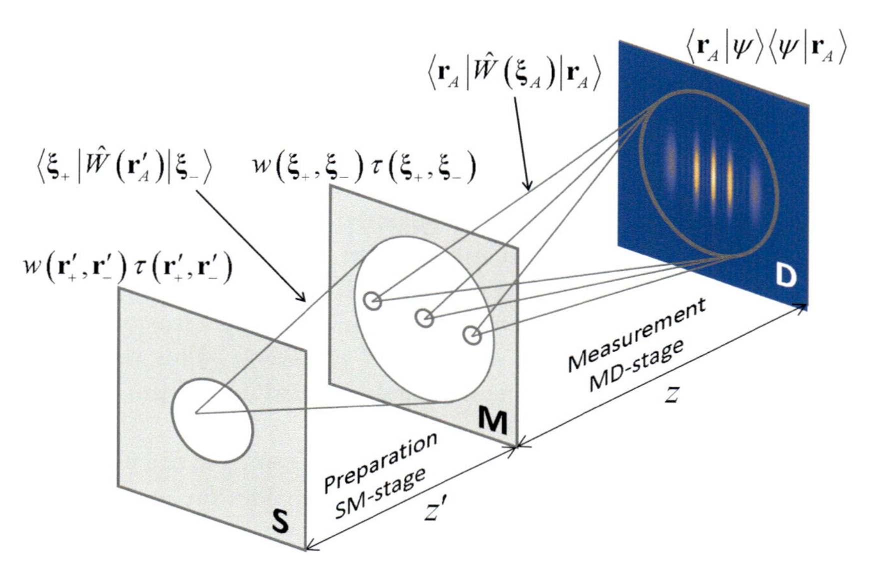

Fig. 5 Conceptual sketch of preparation-and-measurement (P&M) configured interferometers for single photon and single matter particle interference. S, M and D denote the source, mask and detector planes, respectively. The mathematical expressions are explained in the text.

photons or single matter particles is placed at this plane, labelled as S, Fig. 5. This means that $w(\boldsymbol{\xi}_+, \boldsymbol{\xi}_-)$ is essentially of geometric nature.

The configuration of experimental setups depicted in Fig. 5 is called the P&M (preparation and measurement) configuration. Most the basic interferometers used in single-photon and single-matter particle interference are arranged with this configuration.

Once the interferometer is arranged, specific geometric states of space are established in both stages, as described by the quantum non-paraxial operator for interference in vacuum, Eq. (36)

$$|\Upsilon_0\rangle\langle\Upsilon_0| = \frac{\hbar\omega}{2}\begin{cases} \int_S d^2\mathbf{r}'_A \hat{W}(\mathbf{r}'_A) & \text{for the preparation stage} \\ \int_M d^2\xi_A \hat{W}(\boldsymbol{\xi}_A) & \text{for the measurement stage} \end{cases}, \quad (44)$$

where reduced coordinates $(\mathbf{r}'_A, \mathbf{r}'_D)$ are defined for specifying pairs of points $\mathbf{r}'_\pm = \mathbf{r}'_A \pm \mathbf{r}'_D/2$ on S. The goal of the measurement stage is to determine the quantum probability at D, given by the canonical equation for interference in vacuum $\langle \mathbf{r}_A | \Upsilon_0\rangle\langle\Upsilon_0 | \mathbf{r}_A\rangle$, as described before in Section 3.1. Let us concern the preparation stage below.

It is important and opportune to highlight a powerful feature of the quantum non-paraxial operator for interference. Its coordinate representation on individual points of a given plane gives the canonical equation for non-paraxial interference that determines the quantum probability on such plane, while the coordinate representation on pairs of points specifies the non-locality prepared on this plane. It should be noted that this representation includes the canonical equation as the particular case for null separation vectors. Because of its generality, we call the coordinate representation of the quantum operator on pairs of points the *master equation for non-paraxial interference*. This is a crucial feature that supports the generality and exhaustivity of the proposed theory.

Let us regard the quantum operator for interference in the preparation (SM) stage, Fig. 5. The master equation at M provides the prepared non-locality $w(\boldsymbol{\xi}_+, \boldsymbol{\xi}_-) = \left(\frac{2}{\hbar\omega}\right)\langle\boldsymbol{\xi}_+ | \Upsilon_0\rangle\langle\Upsilon_0 | \boldsymbol{\xi}_-\rangle$ at pairs of points $\boldsymbol{\xi}_\pm$ of M, while its local component for $\boldsymbol{\xi}_D = 0$, i.e. the canonical equation, determines the quantum probability $|\psi(\boldsymbol{\xi}_A)|^2 = \left(\frac{2}{\hbar\omega}\right)\langle\boldsymbol{\xi}_A | \Upsilon_0\rangle\langle\Upsilon_0 | \boldsymbol{\xi}_A\rangle$ for a single photon or single matter particle to arrive at a given point $\boldsymbol{\xi}_A$. The area of the source placed at S determines the emitting points, but only one of them is activated in any individual experimental realization. It means that the geometric potential in SM-stage nullifies and

$$\hat{W}(\mathbf{r}'_A) = \hat{R}(\mathbf{r}'_A) = \kappa(\mathbf{r}'_A, \mathbf{r}'_A) \quad |\Theta(\mathbf{r}'_A)\rangle\langle\Theta(\mathbf{r}'_A)|, \tag{45}$$

denotes the ground states in SM stage with vertices at the emitting points of the source, where

$$\kappa(\mathbf{r}'_A, \mathbf{r}'_A) = \begin{cases} |\psi(\mathbf{r}'_A)|^2 \, |t(\mathbf{r}'_A)|^2 & \text{for photons} \\ \cos^2(\beta(\mathbf{r}'_A)) & \\ |\psi(\mathbf{r}'_A)|^2 \, |t(\mathbf{r}'_A)|^2 & \text{for matter particles} \end{cases} \tag{46}$$

determines the quantum probability of emission at the source point $\mathbf{r}'_A$, and $|\Theta(\mathbf{r}'_A)\rangle\langle\Theta(\mathbf{r}'_A)|$ is the density matrix of ground states of space in

SM-stage. So, non-locality cones in SM-stage are given by $\langle \boldsymbol{\xi}_+ | \hat{R}(\mathbf{r}'_A) | \boldsymbol{\xi}_- \rangle = \kappa(\mathbf{r}'_A, \mathbf{r}'_A) \ \langle \boldsymbol{\xi}_+ | \Theta(\mathbf{r}'_A) \rangle \langle \Theta(\mathbf{r}'_A) | \boldsymbol{\xi}_- \rangle$, with

$$\begin{aligned} \langle \boldsymbol{\xi}_+ | \Theta(\mathbf{r}'_A) \rangle \langle \Theta(\mathbf{r}'_A) | \boldsymbol{\xi}_- \rangle &= \left(\frac{k}{4\pi}\right)^2 \left(\frac{z' + |\mathbf{z}' + \boldsymbol{\xi}_+ - \mathbf{r}'_A|}{|\mathbf{z}' + \boldsymbol{\xi}_+ - \mathbf{r}'_A|^2}\right) \\ \times \left(\frac{z' + |\mathbf{z}' + \boldsymbol{\xi}_- - \mathbf{r}'_A|}{|\mathbf{z}' + \boldsymbol{\xi}_- - \mathbf{r}'_A|^2}\right) & \exp(ik|\mathbf{z}' + \boldsymbol{\xi}_+ - \mathbf{r}'_A| - ik|\mathbf{z}' + \boldsymbol{\xi}_- - \mathbf{r}'_A|). \end{aligned} \tag{47}$$

The non-paraxial modes in Eq. (47) spatially modulate to each other as they overlap in Eq. (44), thus producing spatially structured non-locality cones. This implies that $\langle \boldsymbol{\xi}_+ | \hat{R}(\mathbf{r}'_A) | \boldsymbol{\xi}_- \rangle$ denotes a set of non-separable cones, and therefore they must be overlapped for all the individual experimental realizations, in spite that only one emitting point is activated in each case. Accordingly, the prepared non-locality at M is the cross-section of the overlapped cones at this plane, whose area determines the non-locality support (Castañeda et al., 2020). This prepared non-locality is given by the master equation of interference in vacuum

$$\langle \boldsymbol{\xi}_+ | \Upsilon_0 \rangle \langle \Upsilon_0 | \boldsymbol{\xi}_- \rangle = \frac{\hbar\omega}{2} \int_S d^2\mathbf{r}'_A \ \langle \boldsymbol{\xi}_+ | \hat{R}(\mathbf{r}'_A) | \boldsymbol{\xi}_- \rangle, \tag{48}$$

thus indicating that the preparation of non-locality in SM-stage is of geometrical nature. In addition, Eq. (48) establishes that nonlocality at S is not required to prepare nonlocality at M. Actually, a single ground state of space is enough to prepare nonlocality at M. It provides a Lorentzian cone with a maximum on the axis and a monotonic and asymptotic decay to null for points far from the cone axis. Its non-locality support at M is the largest one and is subtended by the angular aperture of the cone (i.e. ca. 80° with respect to the cone axis for a decay of ca. 95 %) (Castañeda et al., 2020). Interesting examples of spatially structured cones for non-locality preparation in SM-stage are reported in (Castañeda et al., 2023).

Now, let us consider the canonical equation of interference at M in any individual experimental realization. It contains only the Lorentzian well with vertex at the activated emitting point, Fig. 6,

$$\begin{aligned} \langle \boldsymbol{\xi}_A | \Upsilon_0 \rangle \langle \Upsilon_0 | \boldsymbol{\xi}_A \rangle &= \frac{\hbar\omega}{2} \ \langle \boldsymbol{\xi}_A | \hat{R}(\mathbf{r}'_A) | \boldsymbol{\xi}_A \rangle \\ &= \frac{\hbar\omega}{2} \kappa(\mathbf{r}'_A, \mathbf{r}'_A) \ \langle \boldsymbol{\xi}_A | \Theta(\mathbf{r}'_A) \rangle \langle \Theta(\mathbf{r}'_A) | \boldsymbol{\xi}_A \rangle, \end{aligned} \tag{49}$$

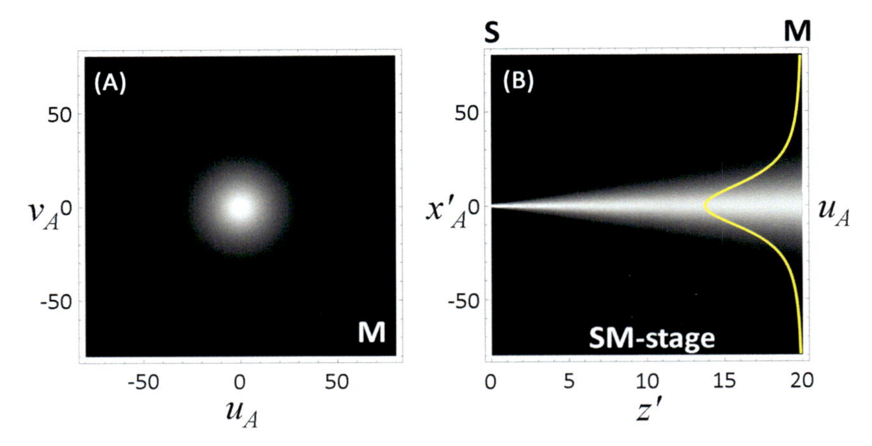

Fig. 6 Lorentzian well in SM-stage of Fig. 5 for the ground state of space with vertex at $\mathbf{r}'_A = 0$ of S, with $\mathbf{r}'_A \equiv (x'_A, y'_A)$. (A) Cross-section on M, at $z' = 50\lambda$ ($\lambda = 0.4\,\mu m$ for single photons and $\lambda = 0.4\,pm$ for single matter particles) describing the quantum probability of arrivals at $\boldsymbol{\xi}_A \equiv (u_A, v_A)$. (B) Axial section in SM-stage for $0 \leq z' \leq 50\lambda$. The vertical Lorentzian profile describes the quantum probability for arrivals at M. Axes units are μm for single photons and pm for single matter particles.

with

$$\langle \boldsymbol{\xi}_A | \Theta(\mathbf{r}'_A) \rangle \langle \Theta(\mathbf{r}'_A) | \boldsymbol{\xi}_A \rangle = \left(\frac{k}{4\pi} \right)^2 \left(\frac{z' + |\mathbf{z}' + \boldsymbol{\xi}_A - \mathbf{r}'_A|}{|\mathbf{z}' + \boldsymbol{\xi}_A - \mathbf{r}'_A|^2} \right)^2. \tag{50}$$

Therefore, $\langle \boldsymbol{\xi}_A | \hat{R}(\mathbf{r}'_A) | \boldsymbol{\xi}_A \rangle$ gives the quantum probability for the emitted photon or matter particle at $\mathbf{r}'_A$ of S to arrive at any $\boldsymbol{\xi}_A$ of M, Fig. 6.

3.3 The individual experimental realization

Events of value 1 or individual experimental realizations in interference experiments with single photons and single matter particles are characterized by $\langle 1 | \hat{n} | 1 \rangle = 1$. Each one is described by three local consecutive events, i.e. emission at S, mask crossing at M and detection at D. The term "local" concerns the corpuscular nature of photons and matter particles and is restricted by quantum (Feynman et al., 1965) and geometric uncertainties

(Castañeda et al., 2023). Therefore, the quantum interference operator for the individual experimental realizations is given by

$$\langle 1 | \Upsilon \rangle \langle \Upsilon | 1 \rangle = |\Upsilon_1\rangle\langle\Upsilon_1| = \frac{3}{2}\hbar\omega\,|\psi\rangle\langle\psi| \\ = \frac{3}{2}\hbar\omega \begin{cases} \int_S d^2r'_A\,\hat{W}(\mathbf{r}'_A) & \text{for the preparation stage} \\ \int_M d^2\xi_A\,\hat{W}(\boldsymbol{\xi}_A) & \text{for the measurement stage} \end{cases}. \tag{51}$$

The only difference between Eqs. (44) and (51) is the numerical coefficient, which equals $1/2$ for vacuum or events of value 0, and $3/2$ for individual experimental realizations or events of value 1. The geometric states of space in the setup, represented by the density operator $|\psi\rangle\langle\psi|$, remain invariant. It is consistent with the phenomenological explanation that the single photon or the single matter particle propagates confined in a geometric state of space. This explanation does not include nor imply or require the wave–particle duality and related hypotheses (Feynman et al., 1965). Furthermore, it has the advantages of restoring the inert character of the corpuscular objects (i.e. photons and matter particles) as they propagate in the interferometer, as well as providing a causal explanation of interference in ordinary space supported by fewer premises than the standard quantum formalism.

An important question concerns the coupling of the single photon or matter particle with the geometric state of space. It is determined by k parameter. This parameter has two meanings, that is, a geometrical meaning concerning the shape of the non-local, non-paraxial modes of the density matrix $|\Theta\rangle\langle\Theta|$, and a physical meaning concerning the single photon frequency ω as well as the momentum p of the single matter particle. Specifically, the harmonic factor of the non-local modes confers to k the geometrical meaning $k = 2\pi/\lambda$, with λ the length along which the argument of the harmonic factor evolves in 2π. So, λ determines a metric for the spatial modulations provided by such modes. The physical meaning of k is given by $k = \omega/c$ for photons and $k = p/\hbar$ for matter particles. The geometrical and the physical meanings of k yield the well-known relationships

$$\lambda = \begin{cases} 2\pi c/\omega & \text{for photons} \\ h/p & \text{for matter particles} \end{cases}, \tag{52}$$

that establish the coupling condition between the non-local modes of metric λ and photons of frequency ω as well as matter particles of momentum p. More precisely, in any individual experimental realization, the single photon or the single matter particle enters the SM-stage at the emission point on S and the MD-stage at the crossing point on M. Such points are vertices of geometric states of space with different metrics at a time. So, the single photon or the single matter particle couples only the geometric state of space excited with metric λ that fulfils the condition in Eq. (52). In case that the setup does not select such geometric state, the single photon or the single matter particle couples the corresponding ground state of space. This coupling mechanism is consistent with the requirement of the appropriate scale in designing similar configured interferometers for single photons and single matter particles.

The phenomenological description of any individual experimental realization is as follows. Once the single photon or single matter particle is emitted at the source point $\mathbf{r}'_0$, with a quantum probability specified by $\kappa(\mathbf{r}'_0, \mathbf{r}'_0)$, it accesses only the ground state of space with vertex at the emission point, $\hat{W}(\mathbf{r}'_0) = \hat{R}(\mathbf{r}'_0)$, with the above mentioned uncertainty restrictions. Then, the single photon or single matter particle moves confined in the Lorentzian well, preferably near its axis, thus arriving at any point $\boldsymbol{\xi}_A$ of M in accordance with the canonical equation of interference

$$\langle \boldsymbol{\xi}_A | \Upsilon_1 \rangle \langle \Upsilon_1 | \boldsymbol{\xi}_A \rangle = \frac{3}{2} \hbar\omega \langle \boldsymbol{\xi}_A | \hat{R}(\mathbf{r}'_0) | \boldsymbol{\xi}_A \rangle, \tag{53}$$

where the Lorentzian well $\langle \boldsymbol{\xi}_A | \hat{R}(\mathbf{r}'_0) | \boldsymbol{\xi}_A \rangle$ determines the arrival quantum probability.

If the effective arrival point $\boldsymbol{\xi}_A = \boldsymbol{\xi}_0$ is transparent, then the single photon or single matter particle accesses only the geometric state of space with vertex at such point, $\hat{W}(\boldsymbol{\xi}_0) = \hat{R}(\boldsymbol{\xi}_0) + \frac{1}{2}\hat{G}(\boldsymbol{\xi}_0)$, and moves confined in the spatially structured Lorentzian well of this geometric state of space. The geometric potential that excites this state of space is activated by the prepared non-locality, determined by the master equation of interference

$$\langle \boldsymbol{\xi}_+ | \Upsilon_1 \rangle \langle \Upsilon_1 | \boldsymbol{\xi}_- \rangle = \frac{3}{2} \hbar\omega \int_S d^2 r'_A \langle \boldsymbol{\xi}_+ | \hat{R}(\mathbf{r}'_A) | \boldsymbol{\xi}_- \rangle. \tag{54}$$

Nevertheless, the canonical equation of interference with this geometric state of space

$$\langle \mathbf{r}_A | \Upsilon'_1 \rangle \langle \Upsilon'_1 | \mathbf{r}_A \rangle = \frac{3}{2}\hbar\omega \left(\langle \mathbf{r}_A | \hat{R}(\boldsymbol{\xi}_0) | \mathbf{r}_A \rangle + \frac{1}{2} \langle \mathbf{r}_A | \hat{G}(\boldsymbol{\xi}_0) | \mathbf{r}_A \rangle \right) \tag{55}$$

cannot determine the arrival quantum probability at all points of D, because this spatially structured Lorentzian well has forbidden zones, constituted by the points $\mathbf{r}_A$ at which $\langle \mathbf{r}_A | \hat{W}(\boldsymbol{\xi}_0) | \mathbf{r}_A \rangle < 0$, as discussed in Section 3.1. However, the arrival quantum probability at D can be determined by modifying the geometric potential in accordance with the spatial entanglement between the set of the geometric states of space in MD-stage, a novel phenomenon described in detail in (Castañeda et al., 2023). Because of the spatial entanglement, the negative values of the forbidden zones of the geometric state of space in any specific individual experimental realization are removed by the confinement zones of the remainder geometric states of space, in spite that they do not confine photons or particles.

More specifically, the pair of geometric states of space excited by each geometric potential mode, under high-valued prepared non-locality, become spatially entangled at their forbidden zones. The spatial entanglement modifies both space states at a time, symmetrically with respect to the z-axis. The forbidden zones of each space state are set to null by reducing the confinement of the coincident zone of the other space state. Therefore, the spatially structured Lorentzian well of the modified geometric state of space $\hat{W}'(\boldsymbol{\xi}_0) = \hat{R}(\boldsymbol{\xi}_0) + \frac{1}{2}\hat{G}'(\boldsymbol{\xi}_0)$ accessed by the single photon or single particle fulfils the condition $\langle \mathbf{r}_A | \hat{W}'(\boldsymbol{\xi}_0) | \mathbf{r}_A \rangle \geq 0$. It should be noted that the spatial entanglement essentially modifies the geometric potential, because the ground states of space are unchangeable.

The canonical equation of interference modified by spatial entanglement,

$$\langle \mathbf{r}_A | \Upsilon_1 \rangle \langle \Upsilon_1 | \mathbf{r}_A \rangle = \frac{3}{2}\hbar\omega \langle \mathbf{r}_A | \hat{W}'(\boldsymbol{\xi}_0) | \mathbf{r}_A \rangle = \frac{3}{2}\hbar\omega \; |\psi'(\mathbf{r}_A)|^2, \quad (56)$$

determines the quantum probability for the single photon or single matter particle arrival to D, thus indicating that it actually moves confined in the modified geometric state of space $\hat{W}'(\boldsymbol{\xi}_0)$. Further important details of spatial entanglement are discussed in Section 5.1.

It should be clarified that the propagation within a confinement zone does not imply or is equivalent to the notion of "path" followed by the photon or the matter particle in each setup stage. This notion is irrelevant for describing interference accurately, and is not considered in the proposed theory. On the other hand, the prediction accuracy of experimental outcomes by the quantum probability calculated in the framework of the standard quantum interference formalism is well-established for a large enough number of individual experimental realizations. However, it seems to fail for only one experimental realization, so that such events are usually regarded as pure random processes. In contrast, the analysis above shows that the proposed theory overcomes this limitation by providing the quantum probability for any individual experimental realization.

3.4 Feynman's mystery of quantum mechanics

The proposed theory also solves what Feynman called "the only mystery of quantum mechanics", concerning single-electron Young (double pinhole) interference (Feynman et al., 1965), as explained in detail in (Castañeda & Hurtado, 2024). Fig. 7 summarizes the explanation. Let us suppose the non-locality function $\kappa(\mathbf{a}) = \tau(\mathbf{a})\ w(\mathbf{a})$ over the mask with two pinholes at $\boldsymbol{\xi}_A = \pm\mathbf{a}/2$, with $\boldsymbol{\xi}_D = \mathbf{a}$ and non-null prepared non-locality $w(\mathbf{a})$, in an experiment with a large enough number of individual realizations, that is $\langle n|\hat{n}|n\rangle = n$ single electrons.

If the whole experiment is performed with both pinholes open, then $\tau(\mathbf{a}) \neq 0$ and $\kappa(\mathbf{a}) \neq 0$ for all individual realizations. As a consequence, each single electron enters one of the two accessible excited states of space, modified by spatial entanglement, $\hat{W}'(\pm\mathbf{a}/2) = \hat{R}(\pm\mathbf{a}/2) + \frac{1}{2}\hat{G}'(0)$, and the canonical equation of interference $\langle \mathbf{r}_A|\Upsilon_n\rangle\langle\Upsilon_n|\mathbf{r}_A\rangle = \left(n + \frac{1}{2}\right)\hbar\omega$ $\left(\langle \mathbf{r}_A|\hat{W}'(\mathbf{a}/2)|\mathbf{r}_A\rangle + \langle \mathbf{r}_A|\hat{W}'(-\mathbf{a}/2)|\mathbf{r}_A\rangle\right)$ determines the quantum probability for electron arrivals to D. Experimental evidence has been reported confirming that the shape of the interference pattern formed by the accumulated records of electron arrivals is corresponding to this quantum probability (Bach et al., 2013; Tavabi et al., 2019). It is illustrated by graphs (A) and (B) in Fig. 7.

However, if the experiment is completely realized by closing any of the pinholes, then $\tau(\mathbf{a}) = 0$ and $\kappa(\mathbf{a}) = 0$ for all individual realizations. As a consequence, both the geometric potential and the geometric state of space with vertex at the blocked pinhole are removed, so that only the ground state with vertex at the open pinhole is accessible, i.e. $\hat{W}(\pm\mathbf{a}/2) = \hat{R}(\pm\mathbf{a}/2)$ and the canonical equation of interference reduces to $\langle \mathbf{r}_A|\Upsilon_n\rangle\ \langle\Upsilon_n|\mathbf{r}_A\rangle = \left(n + \frac{1}{2}\right)\hbar\omega\ \ \langle \mathbf{r}_A|\hat{R}(\pm\mathbf{a}/2)|\mathbf{r}_A\rangle$, thus indicating that the quantum probability for electron arrivals to D has a Lorentzian profile without interference modulations. The experimental evidence confirms this prediction (Bach et al., 2013; Tavabi et al., 2019), which is illustrated in graphs (C) and (D) of Fig. 7.

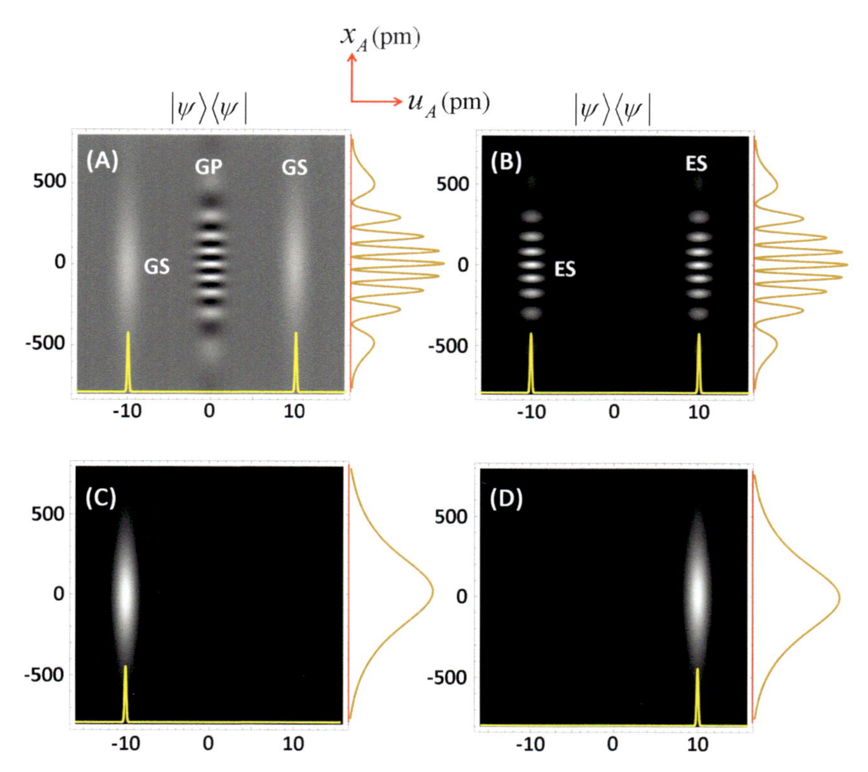

Fig. 7 Density operator $|\psi\rangle\langle\psi|$ of the accessible geometric states of space in Feynman's single electron interference experiment with two pinholes placed at $\boldsymbol{\xi}_A = \pm\mathbf{a}/2$ on M, with $\boldsymbol{\xi}_D = \mathbf{a} = a\,\mathbf{u}_A$ ($a = 5\lambda$, $\lambda = 4\ pm$). The distance between M and D is $z = 100\lambda$. The two pinholes are open in (A), (B), and one of them is blocked in (C), (D). GS: Lorentzian wells of the ground states $\langle \mathbf{r}_A | \hat{R}(\pm\mathbf{a}/2) | \mathbf{r}_A \rangle$ (the Lorentzian profile is appreciable along the vertical white bars on the left and on the right), GP: excitation provided by the geometric potential $\langle \mathbf{r}_A | \hat{G}(0) | \mathbf{r}_A \rangle$ (the cosine-like profile is appreciable along the vertical middle bar), ES: spatially structured Lorentzian wells of the excited states of space $\langle \mathbf{r}_A | \hat{W}'(\pm\mathbf{a}/2) | \mathbf{r}_A \rangle$. The graphs are corresponding to the sections (u_A, x_A) of the geometric space states, determined by mutually parallel components of the vectors $(\boldsymbol{\xi}_A, \mathbf{r}_A)$. The horizontal profiles at the bottom side of each graph describe the quantum probability at M, thus indicating the open pinholes. The vertical profiles on the right side describe the quantum probability at D, thus indicating the pattern recorded by a square modulus detector after large enough number of individual experimental realizations.

4. Interference with classical light

Interference with classical light is described in the well-known quasi-monochromatic wave framework of the optical coherence theory (Born & Wolf, 1993; Mandel & Wolf, 1995). In this framework, it has been established that the waves should be spatially and temporal coherent in order to produce interference patterns by superposition. Usually, temporal coherence is ensured by narrow spectral band light sources, while spatial coherence is determined by the shape and support of the two-point correlation of light, called the cross-spectral density (Born & Wolf, 1993). Furthermore, light is considered as an ergodic, stochastic and stationary electromagnetic wave field (Mandel & Wolf, 1995), so that the cross-spectral density at M is defined as $\overline{\mathbf{E}(\boldsymbol{\xi}_{+})\cdot\mathbf{E}^{*}(\boldsymbol{\xi}_{-})}$, where $\mathbf{E}(\boldsymbol{\xi}_{\pm})$ denotes the electric field vectors of light and the upper bar represents the ensemble average (usually, ensemble average is represented by angular parenthesis $\langle\rangle$, however, the upper bar is used in this paper in order to avoid confusion with Dirac quantum notation).

Nevertheless, an accurate approach to interference with classical light can be achieved in the proposed corpuscular framework for single photon interference. The importance of this approach is the generalization of the proposed theory by including this type of interference.

To this aim, let us consider an extended source that emits a significant large number of photons at a time, $n \gg 1/2$, which do not interact between them. Accordingly, each photon access only one of the geometric states of space in each stage of the setup, in the same way as in the individual experimental realizations of single photon interference. The bosonic nature of photons allows the simultaneous occurrence of the individual experimental realizations, in such a way that the photons can propagate confined at a time in the geometric states of space, and without mutual influences. So, their subsequent recording by the square modulus detector at D is equivalent to the accumulative process of such individual experimental realizations. This phenomenological explanation seems to approach the description of interference with classical light accurately.

It has been shown that the two-point correlation properties of light sources are supported by the non-locality links between pairs of source emitting points (Born & Wolf, 1993; Mandel & Wolf, 1995). This means that excited states of space in SM-stage can be accessible too. More precisely, thermal light sources are essentially spatially incoherent, so that the ground states of space with vertices at each emission point are the only accessible space states for photons emitted at a time. In contrast, laser sources are highly spatially coherent, so that the photons access the excited states of space with vertices at each emission point.

It should be emphasized that, although the optical coherence theory considers non-locality as a physical and statistical attribute of the source (Born & Wolf, 1993; Mandel & Wolf, 1995), we have shown in the preceding sections that non-locality seems to be a geometric attribute of ordinary space described by the quantum interference operator in vacuum. However, the mathematical formulations of both theories can be reconciled by regarding that the spatial correlation of the light source in a specific interference experiment is described by the same mathematical function as the non-locality at S of the corresponding quantum interference operator in vacuum. This assumption is reasonable by regarding that the complex probability amplitude at S, $\psi(\mathbf{r}'_A)$ is completely realized by the photon emissions at a time over the source area, and should be corresponding to the classical wave-front emitted by the source. Therefore, its spatial correlation $\overline{\psi(\mathbf{r}'_+)\psi^*(\mathbf{r}'_-)} = \overline{w(\mathbf{r}'_+, \mathbf{r}'_-)}$ should be corresponding with the cross-spectral density of the classical light at S. Accordingly, the non-locality at S, $w(\mathbf{r}'_+, \mathbf{r}'_-)$, activates the geometric potential with density matrix $|\Theta(\mathbf{r}'_+)\rangle\langle\Theta(\mathbf{r}'_-)|$, whose non-local modes in the SM-stage are given by

$$\begin{aligned}
&\langle \boldsymbol{\xi}_+ | \Theta(\mathbf{r}'_+)\rangle\langle\Theta(\mathbf{r}'_-) | \boldsymbol{\xi}_-\rangle \\
&= \left(\frac{k}{4\pi}\right)^2 \left(\frac{z' + |\mathbf{z}' + \boldsymbol{\xi}_+ - \mathbf{r}'_+|}{|\mathbf{z}' + \boldsymbol{\xi}_+ - \mathbf{r}'_+|^2}\right) \left(\frac{z' + |\mathbf{z}' + \boldsymbol{\xi}_- - \mathbf{r}'_-|}{|\mathbf{z}' + \boldsymbol{\xi}_- - \mathbf{r}'_-|^2}\right) \\
&\quad \times \exp(ik|\mathbf{z}' + \boldsymbol{\xi}_+ - \mathbf{r}'_+| - ik|\mathbf{z}' + \boldsymbol{\xi}_- - \mathbf{r}'_-|)
\end{aligned} \tag{57}$$

and establish excited geometric states of space in this stage. Therefore, the quantum interference operator in vacuum takes the same mathematical form in both setup stages. For this reason, let us denote the reduced coordinates as

$$(\mathbf{q}_A, \mathbf{q}_D) \equiv \begin{cases} (\mathbf{r}'_A, \mathbf{r}'_D) & \text{for S plane in SM} - \text{stage} \\ (\boldsymbol{\xi}_A, \boldsymbol{\xi}_D) & \text{for M plane in MD} - \text{stage} \end{cases}$$

and

$$(\mathbf{u}_A, \mathbf{u}_D) \equiv \begin{cases} (\boldsymbol{\xi}_A, \boldsymbol{\xi}_D) & \text{for M plane in SM} - \text{stage} \\ (\mathbf{r}_A, \mathbf{r}_D) & \text{for D plane in MD} - \text{stage} \end{cases}$$

in order to simplify the notation below, and express the quantum interference operator in vacuum as

$$|\Upsilon_0\rangle\langle\Upsilon_0| = \frac{1}{2}\hbar\omega \int d^2q_A \left(\hat{R}(\mathbf{q}_A) + \frac{1}{2}\hat{G}(\mathbf{q}_A)\right) \tag{58}$$

for the complete experimental setup. For $n >> 1/2$ simultaneous individual experimental realizations, the quantum interference operator takes the form

$$\overline{|\Upsilon_n\rangle\langle\Upsilon_n|} = n\hbar\omega \int d^2q_A \left(\overline{\hat{R}(\mathbf{q}_A) + \frac{1}{2}\hat{G}(\mathbf{q}_A)}\right), \tag{59}$$

with

$$\overline{\hat{R}(\mathbf{q}_A)} = \overline{\kappa(\mathbf{q}_A, \mathbf{q}_A)} \; |\Theta(\mathbf{q}_A)\rangle\langle\Theta(\mathbf{q}_A)| \tag{60}$$

and

$$\overline{\hat{G}(\mathbf{q}_A)} = \int_{\mathbf{q}_D \neq 0} d^2q_D \; \overline{\kappa(\mathbf{q}_+, \mathbf{q}_-)} \; |\Theta(\mathbf{q}_+)\rangle\langle\Theta(\mathbf{q}_-)|, \tag{61}$$

where

$$\begin{aligned} \overline{\kappa(\mathbf{q}_+, \mathbf{q}_-)} = \overline{w(\mathbf{q}_+, \mathbf{q}_-)} \; \tau(\mathbf{q}_+, \mathbf{q}_-) \; \overline{\cos(\beta(\mathbf{q}_+)) \; \cos(\beta(\mathbf{q}_-))} \\ \cos(\theta(\mathbf{q}_+) - \theta(\mathbf{q}_-)), \end{aligned} \tag{62}$$

whose local component is $\overline{\kappa(\mathbf{q}_A, \mathbf{q}_A)} = \overline{|\psi(\mathbf{q}_A)|^2} \; |t(\mathbf{q}_A)|^2 \; \overline{\cos^2(\beta(\mathbf{q}_A))}$.

The factor $n\hbar\omega \; \overline{\kappa(\mathbf{q}_A, \mathbf{q}_A)}$ has an important meaning. The irradiance of classical light is defined as the energy per unit area and unit time (Born & Wolf, 1993), and it is specified by the number of photons found at a time in the vicinity of a point $\mathbf{q}_A$ (Saleh & Teich, 2019). So, $n\hbar\omega$ denotes the light energy provided by n photons at a time, and the irradiance at any point $\mathbf{q}_A$ is determined by the fraction of photons $n \; \overline{|\psi(\mathbf{q}_A)|^2} \; |t(\mathbf{q}_A)|^2 \; \overline{\cos^2(\beta(\mathbf{q}_A))}$. These photons enter the geometric state of space $\overline{\hat{W}(\mathbf{q}_A)} = \overline{\hat{R}(\mathbf{q}_A)} + \frac{1}{2}\overline{\hat{G}(\mathbf{q}_A)}$ at its vertex $\mathbf{q}_A$.

In Eq. (62), the non-local transmission $\tau(\mathbf{q}_+, \mathbf{q}_-)$ and the angle $\theta(\mathbf{q}_+) - \theta(\mathbf{q}_-)$ between the transmission axes at the points $\mathbf{q}_\pm$ are deterministic variables, while the factors $\cos(\beta(\mathbf{q}_\pm))$ are spatially correlated in some extent (Castañeda et al., 2021), i.e. $\overline{\cos(\beta(\mathbf{q}_+))\ \cos(\beta(\mathbf{q}_-))} \neq 0$, even when the angles $\beta(\mathbf{q}_\pm)$ are mutually independent statistical variables, because of the simultaneous mutually parallel components of the polarization states $|\,||_\pm\rangle$.

Angles $\beta(\mathbf{q}_\pm)$ are deterministic for polarized light, so that $\overline{\cos(\beta(\mathbf{q}_+))\ \cos(\beta(\mathbf{q}_-))} = \cos(\beta(\mathbf{q}_+))\ \cos(\beta(\mathbf{q}_-))$, with $0 \leq |\cos(\beta(\mathbf{q}_+))\ \cos(\beta(\mathbf{q}_-))| \leq \cos^2\beta \leq 1$. The null value is achieved for $\beta(\mathbf{q}_\pm) = \pm\pi/2$ and the unit value is achieved for $\beta(\mathbf{q}_\pm) = 0, \pm\pi$, which includes $\beta(\mathbf{q}_+) = \beta(\mathbf{q}_-) = \beta$. For non-polarized light, only the condition $\beta(\mathbf{q}_+) = \beta(\mathbf{q}_-) = \beta$, with β a stochastic variable uniformly distributed on its domain, contributes to the correlation. Consequently, $\overline{\cos(\beta(\mathbf{q}_+))\ \cos(\beta(\mathbf{q}_-))} = \overline{\cos^2\beta} = 1/2$. This means that interference patterns can be recorded with non-polarized classical light, provided it is spatially correlated in some extent, as known since the original Young's experiment (Young, 1804).

Nevertheless, $\overline{\cos(\beta(\mathbf{q}_+))\ \cos(\beta(\mathbf{q}_-))}$ and the prepared cross-spectral density, $\overline{w(\mathbf{q}_+, \mathbf{q}_-)} = \overline{\psi(\mathbf{q}_+)\ \psi^*(\mathbf{q}_-)}$ are statistically independent to each other, such that the corresponding ensemble averages are factorable in Eq. (62).

The cross-spectral densities at the output planes of each stage are given by

$$\begin{aligned}\overline{w(\mathbf{u}_+, \mathbf{u}_-)} &= \frac{1}{n\hbar\omega}\overline{\langle \mathbf{u}_+ | \Upsilon_n\rangle\langle\Upsilon_n | \mathbf{u}_-\rangle} \\ &= \int d^2q_A\left(\overline{\langle \mathbf{u}_+ | \hat{R}(\mathbf{q}_A) | \mathbf{u}_-\rangle + \frac{1}{2}\langle \mathbf{u}_+ | \hat{G}(\mathbf{q}_A) | \mathbf{u}_-\rangle}\right),\end{aligned} \tag{63}$$

whose local component for $\mathbf{u}_D = 0$ determines the corresponding irradiance distributions

$$\begin{aligned}\overline{|\Upsilon_n(\mathbf{u}_A)|^2} &= \overline{\langle \mathbf{u}_A | \Upsilon_n\rangle\langle\Upsilon_n | \mathbf{u}_A\rangle} \\ &= n\hbar\omega \int d^2q_A\left(\overline{\langle \mathbf{u}_A | \hat{R}(\mathbf{q}_A) | \mathbf{u}_A\rangle + \frac{1}{2}\langle \mathbf{u}_A | \hat{G}(\mathbf{q}_A) | \mathbf{u}_A\rangle}\right)\end{aligned} \tag{64}$$

Let us consider some cases of special interest in interference with classical light. Let us start by describing the SM-stage when a spatially incoherent light source is placed at S, such that $\overline{w(\mathbf{q}_+, \mathbf{q}_-)} = 0$, and therefore $\overline{\kappa(\mathbf{q}_+, \mathbf{q}_-)} = 0$ for any $\mathbf{q}_D \neq 0$. Thus, Eq. (63) gives

$$\frac{1}{n\hbar\omega}\overline{\langle \mathbf{u}_+ | \Upsilon_n \rangle \langle \Upsilon_n | \mathbf{u}_- \rangle} = \overline{w(\mathbf{u}_+, \mathbf{u}_-)} = \int d^2 q_A \overline{\langle \mathbf{u}_+ | \hat{R}(\mathbf{q}_A) | \mathbf{u}_- \rangle}$$
$$= \int d^2 q_A \ \overline{|\psi(\mathbf{q}_A)|^2} \ |t(\mathbf{q}_A)|^2 \ \overline{\cos^2(\beta(\mathbf{q}_A))} \langle \mathbf{u}_+ | \Theta(\mathbf{q}_A) \rangle \langle \Theta(\mathbf{q}_A) | \mathbf{u}_- \rangle \tag{65}$$

for the cross-spectral density prepared at M, that involves only the ground states of space in the SM-stage. So, it is also fulfilled by classical light that non-locality at S is not required to prepare non-locality at M. Although Eq. (65) mathematically corresponds to van Cittert–Zernike theorem (Born & Wolf, 1993; Mandel & Wolf, 1995), its phenomenological description in the proposed framework provides a new interpretation of this fundamental theorem of optical coherence theory. Specifically, the cross-spectral density at M is provided by the spatially structured non-locality cone due to the overlapping of the ground states of space in SM-stage, instead of resulting from the gain in spatial coherence of light wave-fronts along their propagation. The advantage of our interpretation is that it does not require the assumption of attributes for the propagation in order to explain the non-locality preparation. In fact, all the emitted photons at a time access only the ground states of space in this stage, thus moving confined in the corresponding Lorentzian wells, without interaction between them. Accordingly, Eq. (64) describes the irradiance distribution that illuminates M as

$$\begin{aligned}\overline{|\Upsilon_n(\mathbf{u}_A)|^2} &= n\hbar\omega \int d^2 q_A \ \overline{\langle \mathbf{u}_A | \hat{R}(\mathbf{q}_A) | \mathbf{u}_A \rangle} \\ &= n\hbar\omega \int d^2 q_A \ \overline{|\psi(\mathbf{q}_A)|^2} \\ &\quad |t(\mathbf{q}_A)|^2 \ \overline{\cos^2(\beta(\mathbf{q}_A))} \\ &\quad \langle \mathbf{u}_A | \Theta(\mathbf{q}_A) \rangle \langle \Theta(\mathbf{q}_A) | \mathbf{u}_A \rangle\end{aligned} \tag{66}$$

Now, let us regard the MD-stage under the condition $\overline{\kappa(\mathbf{q}_+, \mathbf{q}_-)} = 0$ for any $\mathbf{q}_D \neq 0$, which can be achieved by a spatially incoherent illumination of M. Then, Eq. (66) describes the irradiance at D, thus indicating that there is no interference pattern. Nevertheless, the integrand coefficient $\overline{\cos^2(\beta(\mathbf{q}_A))}$, related to the polarization state at M, resembles the classical Malus' law of polarization (Born & Wolf, 1993). Indeed, if the statistical fluctuations of polarization are removed, then the angle $\beta(\mathbf{q}_A)$ becomes deterministic, so that this coefficient takes the form $0 \leq \cos^2(\beta(\mathbf{q}_A)) \leq 1$, with the value null for $\beta(\mathbf{q}_A) = \pm\pi/2$, i.e. light irradiance is absorbed by the polarizer at $\mathbf{q}_A$ because of the orthogonality between the polarization state and the polarizer transmission axis. If this condition is fulfilled over the complete input plane, then $|\Upsilon_n(\mathbf{u}_A)|^2 = 0$ at the output plane. Furthermore, $\cos^2(\beta(\mathbf{q}_A)) = 1$ for $\beta(\mathbf{q}_A) = 0, \pm\pi$, i.e. light irradiance is entirely transmitted by the polarizer at $\mathbf{q}_A$ because the polarization state and the polarizer transmission axis are mutually parallel. If this condition is fulfilled over the complete input plane, then $|\Upsilon_n(\mathbf{u}_A)|^2$ takes on its maximum value at each point of the output plane. For $\beta(\mathbf{q}_A) \neq 0, \ \pm\pi/2, \ \pm\pi$, then $0 < \cos^2(\beta(\mathbf{q}_A)) < 1$, and $|\Upsilon_n(\mathbf{u}_A)|^2$ fulfils Malus' law.

By non-polarized light, $\beta(\mathbf{q}_A)$ is a stochastic variable uniformly distributed over its domain, such that $\overline{\cos^2(\beta(\mathbf{q}_A))} = 1/2$. If the complete input plane is illuminated with non-polarized light and the transmission axes of the polarizers at all points $\mathbf{q}_A$ are mutually parallel, then the well-known result is obtained that the irradiance $|\Upsilon_n(\mathbf{u}_A)|^2$ on the output plane is a half of the irradiance in absence of polarizers (Born & Wolf, 1993), i.e. $|\Upsilon_n(\mathbf{u}_A)|^2 = \frac{n\hbar\omega}{2} \int d^2q_A \ \overline{|\psi(\mathbf{q}_A)|^2} \ |t(\mathbf{q}_A)|^2 \langle \mathbf{u}_A | \Theta(\mathbf{q}_A)\rangle\langle\Theta(\mathbf{q}_A)|\mathbf{u}_A\rangle$.

Let us consider the third case of interest, related to the quantum interference operator in Eq. (59). In this case, the non-locality at M establishes accessible excited states of space in MD-stage. Accordingly, the irradiance at D is described by Eq. (64), with the geometric potential

$$\begin{aligned}
&\overline{\langle \mathbf{u}_A | \hat{G}(\mathbf{q}_A) | \mathbf{u}_A \rangle} \\
&= \int_{\mathbf{q}_D \neq 0} d^2q_D \ \overline{w(\mathbf{q}_+, \mathbf{q}_-)} \ \tau(\mathbf{q}_+, \mathbf{q}_-) \ \overline{\cos(\beta(\mathbf{q}_+)) \ \cos(\beta(\mathbf{q}_-))}. \\
&\times \cos(\theta(\mathbf{q}_+) - \theta(\mathbf{q}_-)) \ \langle \mathbf{u}_A | \Theta(\mathbf{q}_+)\rangle\langle\Theta(\mathbf{q}_-)|\mathbf{u}_A\rangle
\end{aligned} \tag{67}$$

So, Eq. (64) describes interference patterns at D, thus indicating that the non-locality $\overline{\kappa(\mathbf{q}_+, \mathbf{q}_-)}$ at M seems to be the necessary and sufficient condition for interference with classical light too.

The factor $\cos(\theta(\mathbf{q}_+) - \theta(\mathbf{q}_-))$ provided by $\overline{\kappa(\mathbf{q}_+, \mathbf{q}_-)}$ in integrand of Eq. (67) is of particular interest. It is related to the relative orientations of the transmission axes of the polarizers at the points $\mathbf{q}_\pm$, such that its argument takes on values in the interval $[0, \ \pi/2]$. So, the factor nullifies by mutually orthogonal transmission axes, i.e. $\theta(\mathbf{q}_+) - \theta(\mathbf{q}_-) = \pi/2$. Consequently, $\overline{\kappa(\mathbf{q}_+, \mathbf{q}_-)} = 0$ and the corresponding geometric potential mode is not activated, i.e. this mode does not excite the accessible space states with vertices at the points $\mathbf{q}_{\pm}$. Now, $\theta(\mathbf{q}_+) - \theta(\mathbf{q}_-) = 0$ states for mutually parallel transmission axes, so that $\cos(\theta(\mathbf{q}_+) - \theta(\mathbf{q}_-)) = 1$, and $\overline{\kappa(\mathbf{q}_+, \mathbf{q}_-)} = \overline{w(\mathbf{q}_+, \mathbf{q}_-)}\ \tau(\mathbf{q}_+, \mathbf{q}_-)\ \overline{\cos(\beta(\mathbf{q}_+))\ \cos(\beta(\mathbf{q}_-))}$. So, the corresponding geometric potential mode excites the accessible space states with vertices at the points $\mathbf{q}_{\pm}$. For $0 < \theta(\mathbf{q}_+) - \theta(\mathbf{q}_-) < \pi/2$, the mutually parallel components of the transmission axes fulfill Eq. (62) thus activating the corresponding geometric potential mode, too.

Let us apply this analysis to Young's interference with a double pinhole mask at M, with polarizers inserted in the pinholes. Two ground states of space, with vertices at the mask pinholes $\mathbf{q}_{\pm} = \pm\mathbf{a}/2$ are activated by the monomodal geometric potential

$$\begin{aligned}\overline{\langle \mathbf{u}_A | \hat{G}(0) | \mathbf{u}_A \rangle} = \overline{w(\mathbf{a})}\ \tau(\mathbf{a})\ \overline{\cos(\beta(\mathbf{a}/2))\cos(\beta(-\mathbf{a}/2))}\ \cos(\theta(\mathbf{a}/2) \\ - \theta(-\mathbf{a}/2))\ \langle \mathbf{u}_A | \Theta(\mathbf{a}/2) \rangle \langle \Theta(-\mathbf{a}/2) | \mathbf{u}_A \rangle\end{aligned}. \quad (68)$$

For mutually orthogonal transmission axes of the polarizers at the pinholes, Eq. (68) nullifies. The monomodal geometric potential is removed and Eq. (64) takes the form

$$\overline{|\Upsilon_n(\mathbf{u}_A)|^2} = n\hbar\omega\left(\overline{\langle \mathbf{u}_A | \hat{R}(\mathbf{a}/2) | \mathbf{u}_A \rangle + \langle \mathbf{u}_A | \hat{R}(-\mathbf{a}/2) | \mathbf{u}_A \rangle}\right), \quad (69)$$

thus, indicating that only the ground states of space

$$\begin{aligned}\overline{\langle \mathbf{u}_A | \hat{R}(\pm\mathbf{a}/2) | \mathbf{u}_A \rangle} = n\hbar\omega\ \ & \overline{|\psi(\pm\mathbf{a}/2)|^2}\ \ |t(\pm\mathbf{a}/2)|^2 \\ & \overline{\cos^2(\beta(\pm\mathbf{a}/2))}\ \langle \mathbf{u}_A | \Theta(\pm\mathbf{a}/2) \rangle \\ & \langle \Theta(\pm\mathbf{a}/2) | \mathbf{u}_A \rangle\end{aligned}$$

are accessible for the photons. So, the irradiance distribution at D exhibits a Lorentzian profile without interference modulation.

For mutually parallel transmission axes, the monomodal geometric potential in Eq. (68) is activated and excites the two ground states of space, so that Eq. (64) takes the form

$$\overline{|\Upsilon_n(\mathbf{u}_A)|^2} = n\hbar\omega\left(\overline{\langle \mathbf{u}_A | \hat{R}(\mathbf{a}/2) | \mathbf{u}_A \rangle + \langle \mathbf{u}_A | \hat{R}(-\mathbf{a}/2) | \mathbf{u}_A \rangle + \langle \mathbf{u}_A | \hat{G}(0) | \mathbf{u}_A \rangle}\right). \quad (70)$$

This means that excited states of space are accessible for the photons, and the irradiance distribution at D exhibits an interference modulation, providing that angles $\cos(\beta(\pm\mathbf{a}/2))$ are correlated in some extent. If angles $\cos(\beta(\pm\mathbf{a}/2))$ are uncorrelated, the monomodal geometric potential is removed no matter the transmission axes of the polarizers are mutually parallel, and the irradiance distribution at D does not describe an interference pattern.

The phenomenological description above is corresponding to the well-known Fresnel-Arago's classical laws for interference and polarization (Born & Wolf, 1993). Nevertheless, it should be emphasized that these laws were originally formulated inductively from experimental observations, while they are deduced, in the proposed theory, from geometrical conditions established in vacuum for ensuring the accessibility of the geometric states of space to photons. From this point of view, the proposed theory provides a phenomenological support to the empirical classical laws of interference and polarization in a corpuscular quantum framework.

This approach to interference with classical light completes the generality of the quantum interference operator in vacuum, as well as of the confinement in geometric states of space as phenomenological principle for recording interference patterns.

5. Further novel implications

Mathematical description and phenomenological implications of the subjects summarized in this section have been reported in (Castañeda et al., 2020; 2023). These peculiarities of the proposed theory are not considered by the standard quantum interference formalism and give new insight in interference phenomenology. For this reason, a summary of their relevant features is discussed in the following.

5.1 Spatial entanglement

As discussed before, each non-paraxial mode of the geometric potential is associated to a specific pair of points and excites at a time the two ground states of space with vertices at such points. Under high-valued non-locality, the excitation determines specific distributions of confinement zones as well as forbidden zones in each geometric state. As a consequence, these excited states of space become spatially entangled at their forbidden zones, in such a way that the values of the confinement zones of each state reduces in order to remove the coincident forbidden zones of the other state, thus assuring the compliance of the canonical equation of interference in vacuum $\langle \mathbf{u}_A | \Upsilon_0 \rangle \langle \Upsilon_0 | \mathbf{u}_A \rangle = | \Upsilon_0(\mathbf{u}_A) |^2$, which determines the geometric states of space for each individual experimental realization.

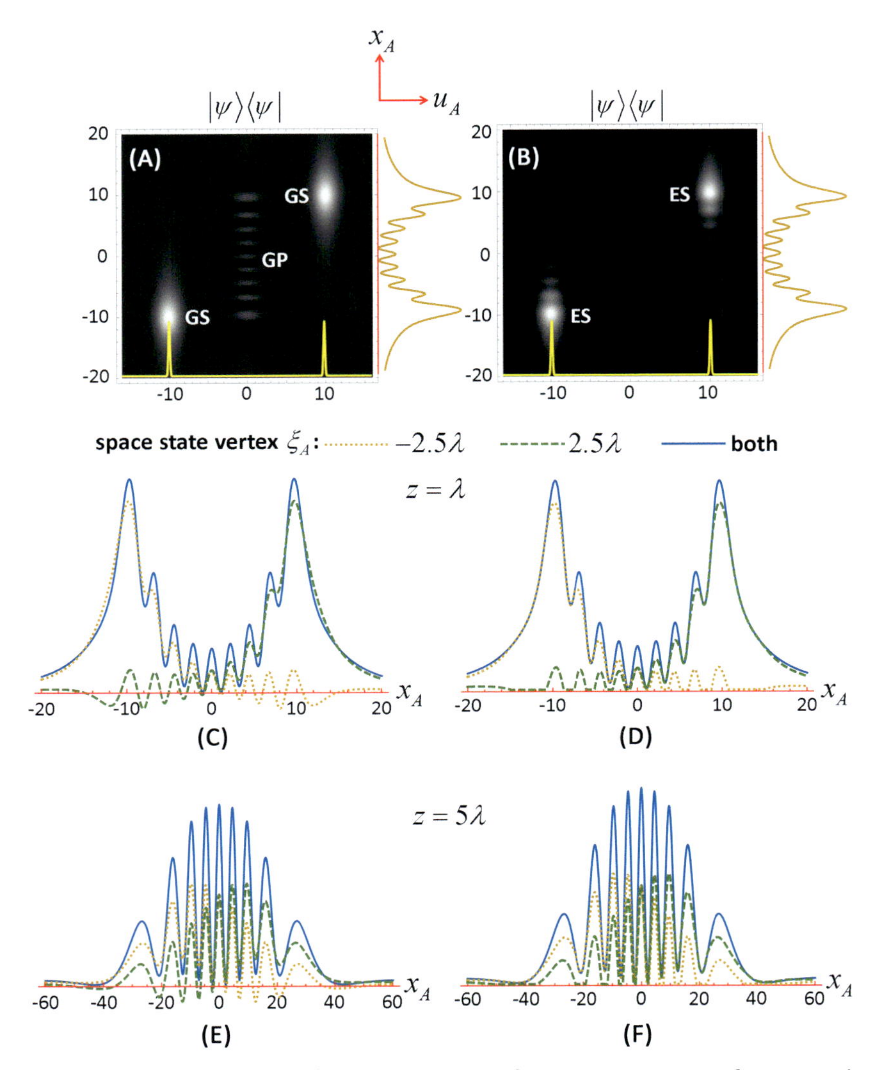

Fig. 8 Spatial entanglement of geometric states of space in Young interference under high-valued prepared non-locality. Pinhole separation vector $\mathbf{a}$ ($|\mathbf{a}| = 5\lambda$, $\lambda = 4\mu m$ for single photons and $4pm$ for single matter particles). Distance from M to D $z = \lambda$. Graphs in (A), B) are sections (u_A, x_A) of the geometric space states, determined by mutually parallel components of the vectors $(\boldsymbol{\xi}_A, \mathbf{r}_A)$. The Lorentzian wells of the two ground states of space (GS, white spots on the left and the right) and the monomodal geometric potential (GP, middle vertical bar) are shown in (A). The spatially structured Lorentzian wells of the excited states of space (GS) are shown in (B). Horizontal profiles on the bottom sides of (A), (B) depict the quantum probability at M, thus indicating the two pinholes. Vertical profiles on the right sides depict the quantum probability at D. Cross-sections profiles of the individual excited states of space (dotted lines) and their

Therefore, Young interference is the most basic experimental situation of spatial entanglement. Let us suppose a mask placed at M, with two pinholes on $\boldsymbol{\xi}_A = \pm\mathbf{a}/2$ and separation vector $\boldsymbol{\xi}_D = \mathbf{a}$, under high-valued prepared non-locality. The monomodal geometric potential excites the ground states of space with vertices at the pinholes, Fig. 8A), thus producing the two excited state of space $\hat{W}(\pm\mathbf{a}/2) = \hat{R}(\pm\mathbf{a}/2) + \frac{1}{2}\hat{G}(0)$ (Fig. 8B). The horizontal profiles on the bottom sides of the graphs depict the quantum probability at M $\langle\boldsymbol{\xi}_A|\psi\rangle\langle\psi|\boldsymbol{\xi}_A\rangle = |\psi(\boldsymbol{\xi}_A)|^2$, while the vertical profiles on the right sides depict the quantum probability at D $\langle\mathbf{r}_A|\psi\rangle\langle\psi|\mathbf{r}_A\rangle = |\psi(\mathbf{r}_A)|^2$.

Confinement and forbidden zones are apparent in the cross-section profiles of the individual geometric states of space in (Fig. 8C and E). It should be noted that the distribution of forbidden zones in each geometric state of space is symmetric to the distribution of forbidden zones in the other one, with respect to the z-axis. So, the spatial entanglement of the two accessible geometric states of space is accurately determined by the violation of the inequality $\langle\mathbf{r}_A|\hat{W}(\mathbf{a}/2)|\mathbf{r}_A\rangle + \langle\mathbf{r}_A|\hat{W}(-\mathbf{a}/2)|\mathbf{r}_A\rangle > \langle\mathbf{r}_A|\hat{W}(\pm\mathbf{a}/2)|\mathbf{r}_A\rangle$. Indeed, it is fulfilled at the points $\mathbf{r}_A$ where confinement zones of both geometric states of space coincide, and violated at the points $\mathbf{r}_A$ where confinement zones of any of the geometric states of space coincide with forbidden zones of the other state.

In the first case, $\langle\mathbf{r}_A|\hat{W}(\pm\mathbf{a}/2)|\mathbf{r}_A\rangle$ determines the quantum probability for the corresponding individual geometric state, but in the second case, $\langle\mathbf{r}_A|\hat{W}(\pm\mathbf{a}/2)|\mathbf{r}_A\rangle$ does not determine such quantum probabilities, although $\langle\mathbf{r}_A|\hat{W}(\mathbf{a}/2)|\mathbf{r}_A\rangle + \langle\mathbf{r}_A|\hat{W}(-\mathbf{a}/2)|\mathbf{r}_A\rangle$ determines the quantum probability at D for the complete experiment. Spatial entanglement of the geometric states of space occurs in these zones with the peculiarity of $0 \leq \langle\mathbf{r}_A|\hat{W}(\mathbf{a}/2)|\mathbf{r}_A\rangle + \langle\mathbf{r}_A|\hat{W}(-\mathbf{a}/2)|\mathbf{r}_A\rangle \leq \langle\mathbf{r}_A|\hat{W}(\pm\mathbf{a}/2)|\mathbf{r}_A\rangle$ because $\langle\mathbf{r}_A|\hat{W}(\mp\mathbf{a}/2)|\mathbf{r}_A\rangle \leq 0$.

Because of the experimental measurement of the individual realization outcomes in interference with single photons and single matter particles

---

overlap (solid line) are shown in (C), (D) for $z = \lambda$ and in (E), (F) for $z = 5\lambda$. Negative valued forbidden zones of the individual geometric states of space are apparent in (C), (E). They are removed by the spatial entanglement of the two geometric states of space in (D), (F). Axes units are pm for single matter particles and μm for single photons.

(Bach et al., 2013; De Martini et al., 1994; Tavabi et al., 2019; Tang, 2022), the quantum probability that predicts such outcomes should be determined. It is feasible by means of a special addition of the individual geometric states of space in the entanglement zones, that set the negative values of the forbidden zones of each geometric state to null by reducing the values of the coincident confinement zones of the other state, as illustrated in Fig. 8D and F. It is equivalent to split the geometric potential into two individualized modes that excite the corresponding ground state of space to $\hat{W}'(\pm\mathbf{a}/2) = \hat{R}(\pm\mathbf{a}/2) + \hat{G}_{\pm}(0)$, with $\hat{G}(0) = \hat{G}_{+}(0) + \hat{G}_{-}(0)$, in such a way that $\langle \mathbf{r}_A | \hat{W}'(\pm\mathbf{a}/2) | \mathbf{r}_A \rangle \geq 0$ determines the quantum probability for the individual experimental realizations, thus restoring the condition $\langle \mathbf{r}_A | \hat{W}'(\mathbf{a}/2) | \mathbf{r}_A \rangle + \langle \mathbf{r}_A | \hat{W}'(-\mathbf{a}/2) | \mathbf{r}_A \rangle > \langle \mathbf{r}_A | \hat{W}'(\pm\mathbf{a}/2) | \mathbf{r}_A \rangle$ at any point of D. This adjustment maintains unchanged the outcomes of the complete experiment predicted by the canonical equation of interference, because the quantum interference operator in vacuum also remains unchanged, i.e.

$$
\begin{aligned}
|\Upsilon_0\rangle \langle \Upsilon_0| &= \frac{\hbar\omega}{2} \left(\hat{W}(\mathbf{a}/2) + \hat{W}(-\mathbf{a}/2)\right) \\
&= \frac{\hbar\omega}{2} \left(\hat{W}'(\mathbf{a}/2) + \hat{W}'(-\mathbf{a}/2)\right).
\end{aligned}
\tag{71}
$$

Summarizing, spatial entanglement seems to be required in order to ensure the measurement of the experimental outcomes of each individual experimental realization of single photon and single matter particle interference. It is an original result of the proposed theory.

It is worth noting that spatial entanglement requires that all geometric states of space live in vacuum in each individual realization, in spite that only one space state confines the single photon or single matter particle. This brings a further important phenomenological implication, i.e. the notion of "collapse" as the automatic removal of the remainder states once a specific state is accessed, is not applicable in the proposed theory. This novel and surprising feature points out that causality in the proposed framework is quite different than in the standard quantum formalism of interference. Specifically, interference is due to the geometric states of space in the proposed framework, and single photons and single matter particles are essentially inert physical objects.

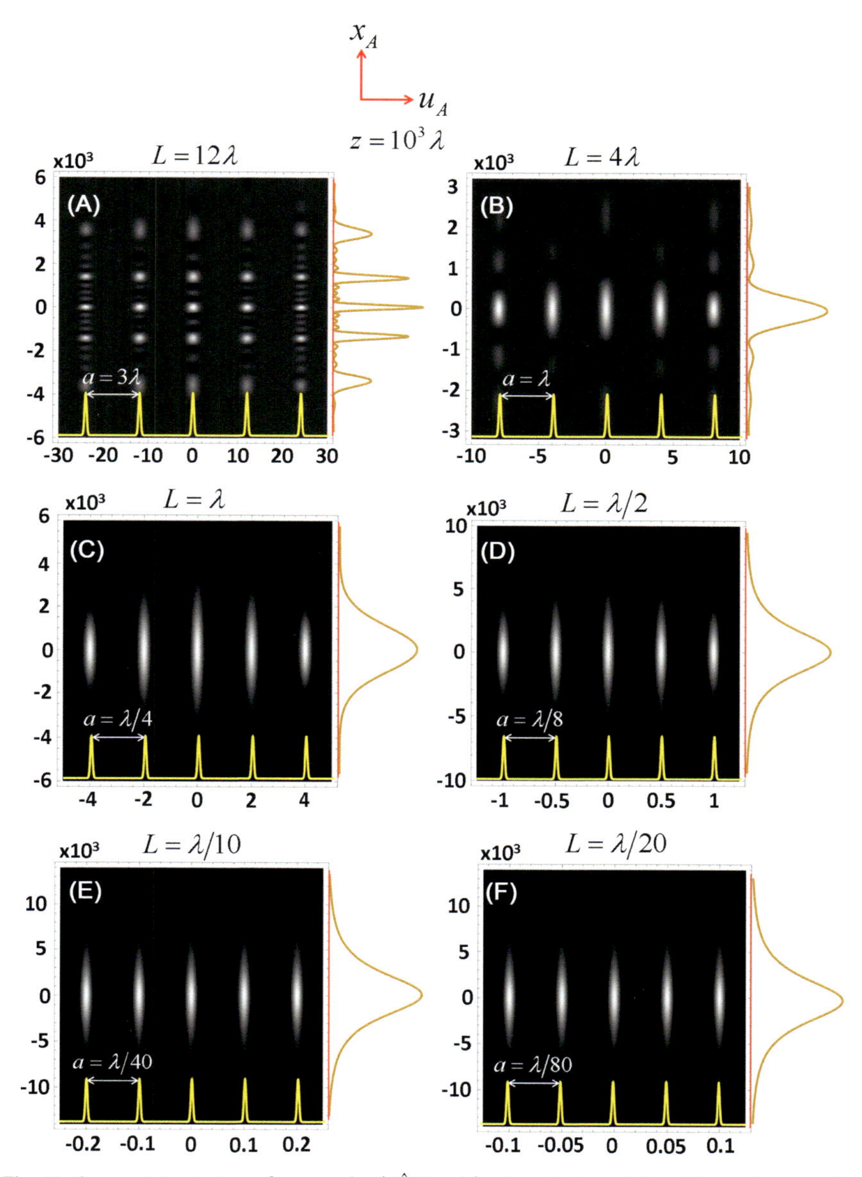

Fig. 9 Geometric states of space $\langle x_A \,|\, \hat{W}(m\ a)\,|\, x_A \rangle$, $-2 \leq m \leq 2$, with vertices at the five pinholes of a grating of spacing a and length $L = 4a$, under high-valued prepared non-locality at M. The distance from M to D is $z = 10^3\lambda$ ($\lambda = 4\ pm$ for single matter particles and $4\ \mu m$ for single photons). The geometric states of space are represented by the five vertical bars in each image, which are corresponding to sections (uA,xA) determined by mutually parallel component of the vectors (ξA,rA). Each state is excited by a specific subset of four geometric potential modes. The excitation fulfils

(Continued)

5.2 Diffraction and geometric uncertainty

It has been shown that the complete distribution of the confinement zones in MD-stage closely depends on the distribution of vertices of the geometric states of space on M (Castañeda et al., 2020). Let us assume that the vertices distribute on 1D array of size L and spacing a on M, i.e. $L \geq a$, and that there is a long enough distance z from M to D. So, it has been established that:

(i) For $a > \lambda$, the geometric potential excites a set of main confinement zones. Consequently, the cross-sections of the excited states of space at D determine a confinement zone pattern in vacuum, whose mapping by the single photons and single matter particles characterizes the interference pattern after a large enough number of individual experimental realizations. It is illustrated in Fig. 9A for a regular linear array of five vertices, i.e. $L = 4a$, under high-valued prepared non-locality.

(ii) For $a \leq \lambda$ and $L > \lambda$, the geometric potential excites only a main confinement zone around the ground-state axes, which is surrounded by significant small secondary confinement zones. Cross-sections of the excited states of space in vacuum determine a pattern at D, whose mapping characterizes the diffraction pattern. It is depicted in Fig. 9B for the same linear array in (i) with a shorter spacing.

(iii) For $(\lambda/10) < L \leq \lambda$, the secondary confinement zones around the central main zone disappear, and the width of the main zone increases. The geometric potential excitation determines only the confinement zone width, which is narrower than the Lorentzian well of a ground

Fig. 9—Cont'd (A) the interference condition, so that each geometric state contributes to the set of main maxima of the interference pattern, (B) the diffraction condition, so that each geometric state contributes to the central main maximum surrounding by small lateral maxima of the diffraction pattern, (C), (D) the condition $\lambda/10 < L \leq \lambda$, and geometric states contributes only to the central main maximum without small lateral maxima. The maximum is narrower than the maximum of the Lorentzian well of a ground state. (E), (F) The geometric uncertainty condition is fulfilled and the geometric states of space are not excited. So, they provide only the Lorentzian wells of the corresponding ground states, whose superposition is a Lorentzian well, too. Horizontal profiles on the bottom sides denote the quantum probability at M, thus describing the pinhole grating. Vertical profiles on the right sides denote the quantum probability at D that predicts the pattern to be recorded after a large enough number of individual experimental realizations. Axes units are *pm* for single matter particles and *μm* for single photons.

state of space. This behavior can be appreciated in Fig. 9C and D by further shortening the spacing of the linear array in (i).

(iv) For $L \leq \lambda/10$, the set of geometric states of space determines only a ground-state Lorentzian well with vertex at any of the array pinholes, no matter the geometric potential, as illustrated in Fig. 9E and F for the linear array of five vertices.

Excellent accuracy of the statements above has been numerically established in (Castañeda et al., 2020). Such statements have the following phenomenological implications:

(i) Diffraction can be described as interference with a finite discrete set of geometric states of space, excited by geometric potential with low spatial frequency modes.

(ii) The geometric potential cannot excite the geometric states of space for $L \leq \lambda/10$, because the angular spreading of the spatial modulations provided by its modes is larger than the angular aperture of the Lorentzian well of the ground states of space. In this case, the result of overlapping the geometric states of space is equivalent to the Lorentzian well of a ground state of space with vertex on any point of the array.

The geometric uncertainty (Castañeda et al., 2023) is a consequence of the last implication. It can be characterized as the indeterminacy of the number of geometric states of space, the indeterminacy of their excitations by the geometric potential and the indeterminacy of the position of their vertices if the array size fulfils the condition $L \leq \lambda/10$. Furthermore, the geometric uncertainty limit $L = \lambda/10$ cannot be reduced by experimental procedures, because it is associated to geometric states of space which live in vacuum.

The analysis above can be extended straightforwardly to arbitrary 2D vertex distributions under any prepared non-locality (Castañeda et al., 2020). It is exemplified in Fig. 10 for a square array of 3×3 vertices. The same array is illustrated in Fig. 4 for $a > \lambda$, under the geometric potential on the middle row. The overlapped geometric state of space for interference is depicted on bottom row.

The array spacing is reduced in Fig. 10 to fit the diffraction condition $a \leq \lambda$ and $L > \lambda$ in (A), (B) . The width of the main confinement zone of the diffraction pattern is clearly narrower than the Lorentzian well of a ground state, with vertex at the array center (dotted line profile). The

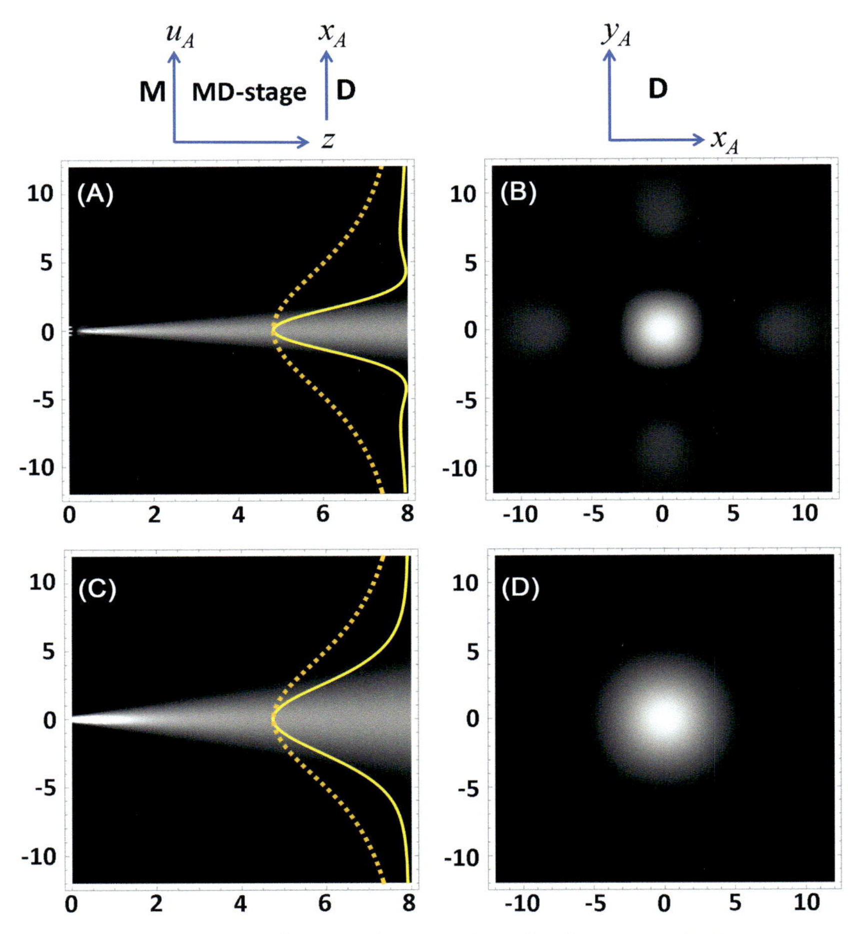

Fig. 10 Geometric states of space whose vertices distribute on a 3 × 3 square array (see Fig. 4G) with horizontal and vertical spacing a and diagonal L, under high-valued prepared nonlocality. Left column: axial sections in MD-stage with $\boldsymbol{\xi}_A \equiv (u_A, v_A)$, $\mathbf{r}_A \equiv (x_A, y_A)$ and $0 \leq z \leq 20\lambda$ ($\lambda = 0.4\ \mu m$ for single photons and $\lambda = 0.4\ pm$ for single matter particles). Right column: cross sections at D ($z = 20\lambda$). Parameters (a, L) take the values $(0.7\lambda, 1.98\lambda)$ in (A), (B), $(0.35\lambda, \lambda)$ in (C), (D), $(0.035\lambda, 0.1\lambda)$ in (E), (F), and $(0.007\lambda, 0.02\lambda)$ in (G), (H). Dotted line profiles on the left column describe the cross-section at D of the ground state Lorentzian well with vertex at the array center. Solid line profiles describe the cross-section at D of the overlapped geometric states of space. They depict a diffraction pattern in (A), and the Lorentzian cross-section of the effective ground state in (E), that remains unchanged in (G). Axes units are μm for single photons and pm for single matter particles.

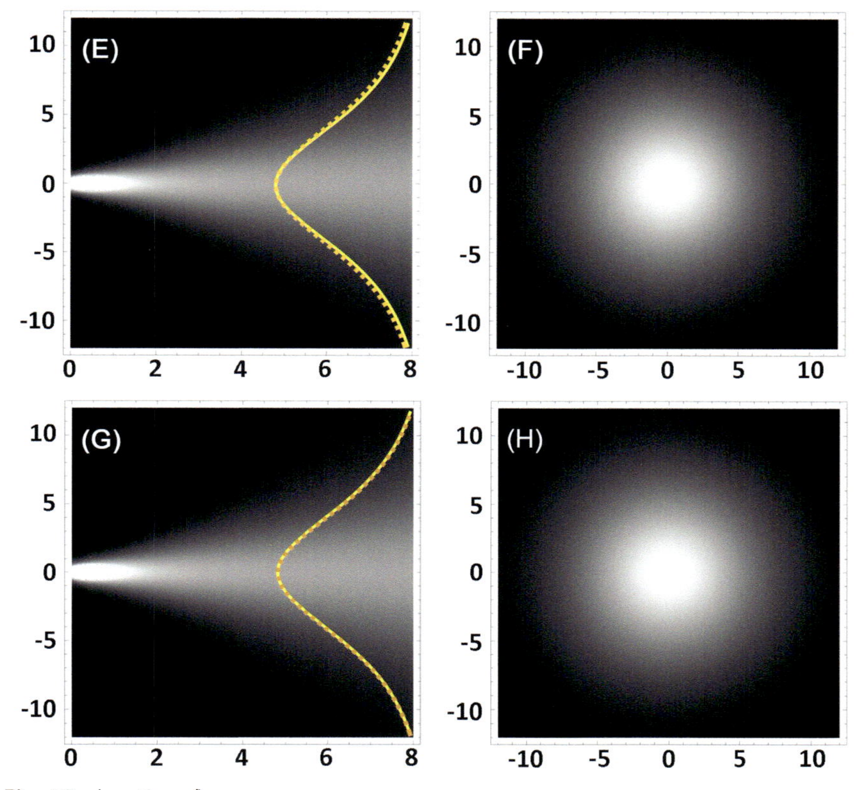

Fig. 10 (*continued*)

condition $(\lambda/10) < L \leq \lambda$ is fitted in (C), (D) so that the main confinement zone of the diffraction pattern is wider than in (a) but remains narrower than the Lorentzian well of the ground state, and the lateral small confinement zones of the pattern in (a) are removed. The geometric uncertainty limit $L = \lambda/10$ is fitted in (E), (F). It is evident that the overlapped geometric state of space becomes identical to the effective ground state. Indeed, the profile of its cross-section at D fits the Lorentzian cross section of the ground state. This behavior remains invariant by further reduction of the array size, i.e. $L < \lambda/10$, as exemplified in (G), (H), thus realizing the geometric uncertainty attributes.

It should be emphasized that geometric uncertainty clearly differs from Heisenberg's uncertainty, which establishes the preclusion of the simultaneous exact determination of the position and momentum of a matter particle. Nevertheless, there is an interesting relationship between geometric uncertainty and Heisenberg's uncertainty, which can be appreciated by considering an individual experimental realization of single matter particle interference by placing a mask with an arbitrary number of pinholes distributed within an area of diameter $L = \lambda/10$. If the pinhole number grows significantly, the array approaches to a continuous aperture. So, Heisenberg's position and momentum uncertainties for the matter particle at M are $\Delta x = L$ and $\Delta p \geq \frac{\hbar}{2L}$, respectively. For $L = \lambda/10$ the minimum momentum uncertainty is $\Delta p_{\min} = 5\frac{\hbar}{\lambda} = \frac{5}{2\pi}p$. It gives the transversal momentum fluctuation $\Delta p_{\min} = p \sin\beta$, where the angle $\beta = \sin^{-1}(5/2\pi) \cong 52.73°$ respective to the axis of the effective ground state.

In turn, the geometric uncertainty establishes that the particle position uncertainty cannot be reduced, i.e. $L = \Delta x_{\min}$ and, because the particle must propagate confined within the Lorentzian well of the effective ground state, then $\Delta p_{\max} = p \sin\beta_0$, with β_0 the aperture angle of the Lorentzian well, which is approximately 70°. Therefore, the particle confinement allows a transversal momentum fluctuation $\Delta p \leq 0.94p$, while Heisenberg's momentum uncertainty establishes a transversal momentum fluctuation $\Delta p \geq 0.79p$.

In other words, the single particle confinement in the effective ground state of space fulfils Heisenberg's momentum uncertainty in the range $0.79p \leq \Delta p \leq 0.94p$. The upper limit of this range is due to the Lorentzian decay of the quantum probability of the particle position at D, determined by the canonical equation $\langle \mathbf{r}_A | \Upsilon_0 \rangle \langle \Upsilon_0 | \mathbf{r}_A \rangle$, which become negligible for $\Delta p \geq 0.94p$. This feature is not reported by the paraxial approximated quantum formalism.

Finally, it is opportune mentioning that the proposed theory is applicable to Michelson interference, too, as discussed in (Bedoya-Rios et al., 2022) for classical light. Such analysis can be straightforwardly extended for single photon and single matter particle interference as follows. The standard description of Michelson interference explains this phenomenon in terms of differences of optical path lengths and involves temporal coherence. Such features are experimentally realized by the lengths of the interferometer arms and the source spectrum, respectively. Indeed, an inverse relationship is well-stablished between the coherence time, associated to the optical path length difference, and the source spectral width (Born & Wolf, 1993).

Coherence time becomes significantly long by sources with narrow spectral widths, i.e. quasi-monochromatic optical sources or quasi-mono-energetic particle sources. It is the case of single photon or single particle interference, in which each individual experimental realization is essentially monochromatic or monoenergetic, respectively. Under this condition,

Michelson interferometers realize the P&M configuration sketched in Fig. 5, by considering a three-dimensional (3D) region around M, determined by the length difference of the interferometer arms. 3D nonlocality supports should be considered, which are in accordance with the 3D geometry of the spatial coherence (Born & Wolf, 1993; Mandel & Wolf, 1995). So, the supports of the prepared non-locality in SM-stage fill the 3D region around M. Furthermore, the vertices of the geometric states of space distribute within such a region, in such a way that the ground states of space are excited with the geometric potential modes with vertices also distributed in the 3D region.

Therefore, the quantum interference operator is given by Eq. (33), too, but now reduced coordinates ($\boldsymbol{\eta}_A$, $\boldsymbol{\eta}_D$) should be introduced to denote pairs of points $\boldsymbol{\eta}_\pm = \boldsymbol{\eta}_A \pm \boldsymbol{\eta}_D/2$ in that 3D region, so that $|\Theta(\boldsymbol{\eta}_A)\rangle\langle\Theta(\boldsymbol{\eta}_A)|$ and $|\Theta(\boldsymbol{\eta}_+)\rangle\langle\Theta(\boldsymbol{\eta}_-)|$ are the density matrices for the ground states and the geometric potential modes that excite them in the Michelson interferometer.

6. Conclusion

The mathematical foundations of a new generalized quantum non-paraxial theory of interference in ordinary space have been discussed, as well as its peculiar phenomenological implications. In comparison with the well-known standard quantum and optical formalisms, this novel theory has clear advantages. The most important one is its generalization. With basis on the same principle of confinement in geometric states of space, it describes in the same way interference with single photons, single matter particles and classical light. In addition, the new theory gives theoretical support to the empirical classical Malus and Fresnel-Arago's laws as well as a new interpretation of the Van Cittert – Zernike theorem. It is also remarkable that the new theory describes interference, exact and exhaustively, in ordinary space-time, thus restoring it as the environment in which quantum interference takes place. Standard quantum interference formalism is mainly predictive and paraxially approximated. Indeed, it determines the paraxial quantum wave function as a vector of Hilbert space without specific physical meaning, and uses it to calculate the quantum probability of experimental outcomes for significant large numbers of photons or matter particles. In contrast, the new theory bases on the quantized Maxwell equations for photons and the Schrödinger equation for matter particles in ordinary space-time, thus regarding the exact (non-paraxial) solution for their spatial components, and the symmetry properties of their time components. In this way, the same set of non-paraxial geometric modes is rigorously deduced for interference with photons and matter particles, and the same dependence from the number of corpuscular objects (single photons or single matter particles) is introduced, each one contributing the fundamental energy $E = \hbar\omega$. This mathematical structure characterizes interference as a stationary (time-independent) and mainly geometric (spatial) phenomenon explicitly.

It should be highlighted that the geometric modes do not disappear as the number of corpuscular objects nullifies. It is not the case of the paraxially approximated standard formalism, in which only the vacuum energy remains under such condition, and this energy is therefore regarded as fluctuations provided by vacuum to the energy of the states of photons or matter particles. That peculiarity of the new theory suggests that vacuum provides much more than energy fluctuations. Indeed, the non-paraxial geometric modes live in vacuum with vacuum energy, and this peculiarity allows proposing that such modes constitute non-paraxial geometric states of ordinary space, which ultimately determine interference by considering that the number of photons or matter particles does not have effects on the interference geometry. So, this context restores the passive (inert) nature of the corpuscular objects and maintains its locality, only restricted by the uncertainty principle. More precisely, in this corpuscular framework interference is not produced by physical or statistical properties of corpuscular objects, so that the wave-particle duality and related hypotheses like delocalization, self-interference and collapse of the wave function are irrelevant. In other words, the new theory provides accurate predictions under fewer assumptions than the standard quantum interference formalism. Furthermore, the new theory is developed in a corpuscular framework that consistently explains what Feynman called the unique mystery of quantum mechanics, thus contributing to the solution of the measurement problem and putting in question, at a time, that interference is a proof of the wave nature. Actually, interference with classical light waves has been presented from a corpuscular point of view, as the result obtained with a significant large number of simultaneous photons.

The non-Newtonian nature of ordinary space is also a disruptive feature of the new theory. A homogeneous, isotropic and passive Newtonian space is considered in the standard quantum and optical interference formalisms.

Consequently, interference is conceived as a peculiar wave behavior, which results challenging to understand interference with particles. Therefore, exotic hypotheses were proposed to reconcile the same phenomenon with mutually exclusive physical objects (i.e. waves and particles). The new theory conceives ordinary space as a system with geometric states, whose ground states are non-locally excited by geometric potentials that modulate them spatially. Such a non-locality seems to be the ultimate cause of interference. This non-Newtonian nature of space, established in vacuum, remains unchanged for any number of single photons or single matter particles.

This is mathematically formalized by the novel quantum interference operator in vacuum, whose coordinate representation for pairs of points in space is the master equation for non-paraxial interference, that determines the non-locality functions which are the necessary and sufficient conditions for interference. The local component of the master equation, i.e. the coordinate representation of the interference operator for single points, is the canonical equation for interference, which describes the geometric states of space in terms of spatially structured Lorentzian wells, thus determining the quantum probability for the measurement of interference patterns. Therefore, an interferometer is a device that filters a specific set of geometric states of the non-Newtonian ordinary space in vacuum. The filtering is performed by non-locality functions established as boundary conditions at specific planes of the device. Therefore, passive corpuscular objects move confined in these geometric states of space, thus mapping the geometry of their spatially structured Lorentzian wells. The map recorded by a square modulus detector after a large enough number of local detection events is called interference pattern. This active role of the device is consistent with basic notions to this respect, early discussed in quantum mechanics, mainly by Bohr.

It should be highlighted that the confinement in geometric states of space is phenomenologically described and mathematically formalized not only for large numbers of photons or matter particles, but also for only one of them in any individual experimental realization. The standard quantum formalism conceives the only-one particle events as pure random that cannot be accurately predicted, because it is not possible to define the quantum probability in an individual experimental realization appropriately due to the negative values involved in its calculation. By considering the spatial entanglement, the new theory overcomes this limitation thus accounting for any individual experimental realization accurately.

Further peculiarities of the new theory were discussed too. Diffraction was presented as low spatial frequency interference. However, spatial entanglement and geometric uncertainty are more relevant features, both closely related with the non-local nature of the non-Newtonian ordinary space. Because of spatial entanglement, the geometric states of space become non-separable, so that non-locality effects between them should be considered in each individual experimental realization, no matter that the single photon or the single matter particle is confined in only one of the space states. The geometric uncertainty determines that any set of geometric states of space whose vertices distribute within an area of size no longer than $\lambda/10$ is equivalent to a unique ground state of space with vertex at any point within such area, no matter the presence of geometric potential. Furthermore, because the geometric uncertainty cannot be reduced, then it establishes a novel upper limit for Heisenberg's uncertainty. These peculiarities are distinctive of the new theory and are not considered in the standard formalism.

The generality, accuracy and exhaustivity of the quantum theory of non-paraxial interference point out that the non-paraxial quantum interference operator is a powerful tool that improves interference understanding in a corpuscular framework and provides theoretical support for interferometer design and control.

Statements and declarations

Ethical compliance

This requirement is not applicable in this study.

Funding

This work was carried out under the research program *Semillero 2241 - Ciencia de la Luz* of the Universidad Nacional de Colombia.

Conflict of interest

The author declare no conflicts of interest.

Data availability

No data were generated or analyzed in the present study.

References

Bach, R., Pope, D., Liou, S.-H., & Batelaan, H. (2013). Controlled double-slit electron diffraction. *New Journal of Physics, 15*, 033018 7pp.

Bedoya-Rios, P., Laverde, J., & Castaneda, R. (2022). Nonparaxial interference and diffraction under 3D spatial coherence. *Journal of the Optical Society of America. A, Optics and Image Science, 39*, 1558.

Bohr, N. (1935). Can quantum-mechanical description of physical reality be considered complete? *Physical Review, 48*, 696–702.

Born, M., & Wolf, E. (1993). *Principles of optics* (sixth ed.). Oxford: Pergamon Press.

Brunner, N., Cavalcanti, D., Pironio, S., Scarani, V., & Wehner, S. (2014). Bell nonlocality. *Reviews of Modern Physics, 86*, 419.

Castañeda, R., Bedoya, P., & Hurtado, C. (2023). Quantum formalism of interference as confinement in spatially structured Lorentzian wells. *Journal of Physics A: Mathematical and Theoretical, 56*, 045302.

Castañeda, R., & Hurtado, C. (2024). Chapter two–The "only mystery of Quantum Mechanics" explained by generalized interference phenomenology. In M. Hÿtch, & P. W. Hawkes (Vol. Eds.), *Advances in imaging and electron physics: 229*, (pp. 43–78). London: Elsevier.

Castañeda, R., Moreno, J., & Colorado, D. (2021). Spatially correlated polarization and nonparaxial propagation of electromagnetic wave fields. *Optical Communication, 481*, 13 126554.

Castañeda, R., Moreno, J., Colorado, D., & Laverde, J. (2020). 3D nonparaxial kernel for two-point correlation modeling in optical and quantum interference at the micro and nanoscales. *Physica Scripta, 95*, 15 065502.

De Martini, F., Denardo, G., & Zeilinger, A. (Eds.). (1994). *Proceedings of the adriatico workshop on quantum interferometry*. Singapore: World Scientific.

Dong-Yeop, N., & Weng Cho, C. (2020). Classical and quantum electromagnetic interferences: What is the difference? *Progress In Electromagnetics Research, 168*, 1–13.

Eberly, J. H., Xiao-Feng, Q., Asma Al, Q., Hazrat, A., Alonso, M. A., Gutiérrez-Cuevas, R., ... Vamivakas, A. N. (2016). Quantum and classical optics–emerging links. *Physica Scripta, 91*, 063003 10pp.

Feynman, R., Leighton, R., & Sands, M. (1965). *The Feynman lectures on physicsVol. 3*. Menlo Park: Addison-Wesley.

Hall, M. J. W. (2016). Chapter eleven–The significance of measurement independence for Bell inequalities and locality. In T. Asselmeyer-Maluga (Ed.). *At the frontier of spacetime* (pp. 189–204). Switzerland: Springer.

Mandel, L., & Wolf, E. (1995). *Optical coherence and quantum optics*. Cambridge: Cambridge University Press.

Saleh, B. E. A., & Teich, M. C. (2019). *Fundamentals of photonics* (third ed.). New York: Wiley.

Scully, M. O., & Zubairy, S. (1997). *Quantum optics*. Cambridge: Cambridge University Press.

Shih, Y. (2020). *An introduction to quantum optics: Photon and biphoton physics*. CRC Press.

Tang, J., & Hu, Z. B. (2022). Analysis of single-photon self-interference in young's double-slit experiments. *The Results in Optics, 9*, 100281.

Tavabi, A. H., Boothroyd, C. B., Yücelen, E., Frabboni, S., Gazzadi, G. C., Dunin-Borkowski, R. E., & Pozzi, G. (2019). The Young-Feynman controlled double-slit electron interference experiment. *Scientific Reports, 9*, 10458.

Walls, D. F., & Milburn, G. J. (1995). *Quantum optics*. Berlin: Springer.

Wódkiewicz, K., & Herling, G. H. (1998). Classical and nonclassical interference. *Physical Review. A, 57*, 815.

Young, T. (1804). The Bakerian lecture: Experiments and calculations relative to physical optics. *Philosophical Transactions of the Royal Society of London, 94*, 1–16.

Zeidler, E. (2011). *Quantum field theory III: Gauge theory, a bridgebetween mathematicians and physicists*. Springer Science and Business Media.

Index

Note: Page numbers followed by "*f*" indicate figures.

H

I

K

L

M

Printed in the United States
by Baker & Taylor Publisher Services